Surface Acoustic Wave Devices

Surface Acoustic Wave Devices

Supriyo Datta

Associate Professor of Electrical Engineering
Purdue University

Prentice-Hall, Englewood Cliffs, NJ 07632

Library of Congress Cataloging-in-Publication Data

Datta, Supriyo, (date)
  Surface acoustic wave devices.

  Bibliography: p.
  Includes index.
  1. Acoustic surface wave devices. I. Title.
TK5981.D35   1986   621.38'0432   85-12110
ISBN 0-13-877911-2

Editorial/production supervision: Sophie Papanikolaou
Cover design: Edsal Enterprises
Manufacturing buyer: Rhett Conklin
Page layout: Gail Collis

© 1986 by Prentice-Hall
A Division of Simon & Schuster, Inc.
Englewood Cliffs, New Jersey 07632

All rights reserved. No part of this book may be
reproduced, in any form or by any means,
without permission in writing from the publisher.

Printed in the United States of America

10 9 8 7 6 5 4 3 2 1

ISBN 0-13-877911-2   025

Prentice-Hall International (UK) Limited, *London*
Prentice-Hall of Australia Pty. Limited, *Sydney*
Editora Prentice-Hall do Brasil, Ltda., *Rio de Janeiro*
Prentice-Hall Canada Inc., *Toronto*
Prentice-Hall Hispanoamericana, S.A., *Mexico*
Prentice-Hall of India Private Limited, *New Delhi*
Prentice-Hall of Japan, Inc., *Tokyo*
Prentice-Hall of Southeast Asia Pte. Ltd., *Singapore*
Whitehall Books Limited, *Wellington, New Zealand*

Dedicated to the memory of my Grandfather

Contents

Preface   xi

Part 1: Background

Chapter 1   *Transversal Filters*   1

1.1   Frequency Response and Impulse Response   *2*

1.2   Some Properties of the Fourier Transform   *7*

1.3   Recursive and Non-Recursive Filters   *14*

1.4   Design of Non-Recursive (Transversal) Filters   *20*

Chapter 2   *Transmission Lines and Plane Acoustic Waves*   29

2.1   Transmission Lines   *30*

2.2 Uniform Plane Acoustic Waves *40*

2.3 Piezoelectricity *52*

2.4 Generation of Acoustic Waves in Piezoelectric Solids *54*

Chapter 3 Surface Acoustic Waves in Piezoelectric Solids 68

3.1 Introductory Description of SAW *69*

3.2 Transmission-Line Model for SAW *72*

3.3 Relation between K^2 and y_0 *76*

3.4 Factors Influencing the Choice of Substrate Cut and Orientation *82*

Part 2: SAW Device Components

Chapter 4 Interdigital Transducers 84

4.1 Definition of Transmitter and Receiver Response Functions: Principle of Reciprocity *85*

4.2 Transmitter Response Function, $\mu(f)$ *91*

4.3 Transducer Admittance *120*

4.4 Model for Numerical Analysis *129*

Chapter 5 Multistrip Couplers 139

5.1 Coupling of Tracks by a Single Electrode *141*

5.2 Overall Coupler Operation *143*

Chapter 6 Reflectors 151

6.1 Operation of a Reflector Array *153*

6.2 Reflection and Velocity Change Due to a Single Electrode *161*

6.3 Reflection of SAW by Electrodes:
     Similarities and Differences with Transmission Lines *169*

Chapter 7 Attenuators and Amplifiers 179

7.1 Electrical Loading *179*

7.2 Mechanical Loading *184*

Chapter 8 Waveguides 187

8.1 Isotropic Substrate *187*

8.2 Anisotropic Substrate *192*

Part 3: SAW Devices

Chapter 9 Bandpass Filters 194

9.1 Capabilities of SAW Bandpass Filters *195*

9.2 Basic Design Procedure *197*

9.3 Analysis *198*

9.4 Second-Order Effects *215*

9.5 Illustrative Examples of Filter Design *218*

Chapter 10 Resonators **225**

10.1   Capabilities of SAW Resonators *226*

10.2   One-Port Resonator *227*

10.3   Two-Port Resonator *235*

10.4   Illustrative Example of Resonator Design *237*

References **240**

    General *240*

    Chapter 1 *241*

    Chapter 2 *242*

    Chapter 3 *242*

    Chapter 4 *243*

    Chapter 5 *243*

    Chapter 6 *244*

    Chapter 7 *245*

    Chapter 8 *245*

    Chapter 9 *245*

    Chapter 10 *246*

Index **247**

Preface

A signal processor must have a memory of its past inputs, since it is impossible to distinguish between two signals on an instantaneous basis. Implementing this memory is generally the costliest part of a signal processor. Conventional *LC* filters derive this memory from a resonant circuit while digital filters use shift registers. Surface acoustic wave (SAW) devices are essentially tapped delay lines that provide a compact, low-power memory, orders of magnitude faster than their digital counterparts. The high-speed memory, together with arithmetic functions implemented on the SAW device, make it attractive for signal processing with bandwidths in the range 10 to 1000 MHz, where the devices are small in size and easily fabricated by photolithographic techniques used in the integrated-circuit industry.

Surface acoustic wave devices are becoming increasingly popular in signal processors in the VHF/UHF range, because of their size and simplicity. However, it is extremely difficult for a beginner with a background in signal processing to understand the operation of SAW devices because of a lack of acquaintance with acoustics. In fact, much of the literature involves a fairly high-level background in acoustic fields and waves that is essential to understanding how energy is coupled into and out of a surface acoustic wave delay line. This proves to be the biggest hurdle to a beginner.

There are several excellent *graduate*-level textbooks such as *Acoustic Fields and Waves in Solids* (B. A. Auld, Wiley, 1973) or *Principles of Acoustic Devices* (V. M. Ristic, Wiley, 1983); advanced books on SAW devices such as *Surface Wave Filters* (ed. H. Matthews, Wiley, 1977) or *Surface Acoustic Waves* (ed. A. A. Oliner, Springer-Verlag, 1978) are also available. However, we felt that in view of the demand for electrical engineers acquainted with SAW device design, an *undergraduate* textbook on the subject would be useful. At the University of Illinois, Bill Hunsinger and I developed a course on SAW devices for seniors who have only an elementary background in signal processing and transmission-line theory. This book grew out of a series of lecture notes developed for this course. We believe that it can be used (1) for undergraduate teaching, (2) for self-study by professionals in the area of signal processing and systems design who wish to acquaint themselves with the terminal properties of SAW devices, and (3) as a reference by engineers actively involved in the design of SAW devices. It has been our goal in this book to develop the essentials of surface wave devices, with a large number of practical examples, relying more on an intuitive approach than on rigorous mathematics. We have made every effort to keep the description at a level that can be followed by an undergraduate student with only an elementary background in signal processing and transmission-line theory. Relevant references are provided from the literature for supplementary reading, although no effort has been made to be exhaustive; the list of references is not meant as a complete bibliography.

The first part of the book is meant to provide a brief introduction to transversal filters and acoustic waves for undergraduate students in electrical engineering who often have no formal acquaintance with either of these subjects. Chapter 1, on transversal filters, is also useful as an introduction to SAW filters, which are basically transversal filters implemented with surface acoustic wave delay lines. Chapter 2 describes the propagation and generation of uniform plane acoustic waves in piezoelectric solids emphasizing the analogy with electrical waves in a transmission line. In Chapter 3 we discuss the surface acoustic wave, which is a nonuniform wave with complicated acoustic and piezoelectric fields coupled together. Throughout the rest of the book we use the *electrostatic potential at the surface* (ϕ) as a measure of the amplitude of the SAW. In the literature, an amplitude for the SAW is usually not defined precisely; this leads to difficulties in problems where the phase information is important. We have chosen the potential as the amplitude (rather than, say, strain) so as to eliminate repeated interconversion between electrical and (often unfamiliar) mechanical quantities. The relationship of ϕ to the power carried by the SAW and to the other field quantities is described in Chapter 3 and a simple transmission line equivalent is developed.

The second part of the book describes the basic components that are used in SAW devices transducers, couplers, and reflectors. Two chapters also include the less common components: attenuators, amplifiers, and waveguides. Our treatment of transducers is slightly different from the usual description found in the literature. In the literature, the radiation conductance (G_a) is determined directly by equating the power dissipated by it to the power carried away by the SAW. We have found it useful to define a transmitter response function (μ) and a receiver response function (g_m) from which the radiation conductance is readily determined. The advantage of these response functions is that unlike G_a, they preserve the phase information and can be used to analyze problems like a transducer in a cavity where the waves running in opposite directions interfere. In Chapter 6 we have included a section that describes how the reflectivity and velocity shift per electrode can be determined from material parameters. These results are fairly recent and have not yet been verified conclusively by experiments. However, they agree with the available experimental data and we felt that the inclusion of these results will be useful despite the lack of complete verification.

The third part of the book describes the design and operation of two different types of SAW devices that have found widespread application: bandpass filters and resonators. Matched filters, convolvers, and correlators have been omitted despite their considerable importance partly because of the diversity of the field and partly because of the author's lack of experience in this area. The systems aspect of SAW devices has also not been discussed. Some of these topics are covered well in *Surface Wave Filters* (ed. H. Matthews, Wiley, 1977) and *Surface Acoustic Waves* (ed. A. A. Oliner, Springer-Verlag, 1978). Only passing references have been made to topics of current research such as unidirectional transducers for low-loss filters or SAW in layered media (such as ZnO on silicon). At the start of each chapter we have summarized the contents and the important tables and results. A large number of examples have been included to help clarify the ideas.

This book would not be written without Bill Hunsinger's active help, guidance, and encouragement; it is difficult for me to express my thanks adequately. Many of the ideas and explanations used in this book were developed through our mutual discussions, and his contributions almost qualify him as a coauthor. I am thankful to Steve Wilkus for his critical comments on the manuscript and his constant help; to Mike Hoskins and Martin Brophy for carefully proofreading the manuscript; to R. J. Kansy of RF Monolithics, Inc., to Tom Martin and Pierre Dufilie of Phonon, Inc., and to John Gau of Anderson Laboratories, Inc., for providing examples of device design and analysis; to S. J. Martin for some of the illustrations; and to Linda Stovall for her skillful typing of the manuscript. I would also like to thank R. Adler, R. L. Gunshor, R. F. Pierret, and numerous other friends and colleagues with whom I have had the pleasure of working with and N. B. Chakraborti, N. R. Sinha Ray, and S. K. Lahiri, who introduced me to the subject. Finally, I would like to thank my wife, Anuradha, and my relatives for their continuing support and encouragement.

Supriyo Datta

Surface Acoustic Wave Devices

1 | Part 1: Background

TRANSVERSAL FILTERS

In 1940, H. E. Kallman proposed the idea of transversal filters (Ref. 1.1), which are radically different from conventional *RLC* filters. A filter discriminates between signals of different frequencies; to do this it needs a memory, since all frequencies look alike at a single instant in time. *RLC* filters derive this memory from a resonant circuit. Transversal filters use a delay line that is tapped along its length. The output is obtained as a weighted sum of the signals at different taps (Fig. 1.10a). SAW filters (Fig. 1.10c) provide a compact implementation of a tapped delay line. Acoustic waves with their relatively slow velocity (~3500 m/s) produce a significant delay within a convenient length; also, the energy is readily tapped from a surface wave because it resides within a wavelength of the surface.

The purpose of this chapter is to give a brief introduction to the analysis and design of transversal filters. Since transversal filters are specified in the frequency domain but are designed in the time domain, it is important to understand the relationship between the frequency response and the impulse response and the properties of the Fourier transform. These are discussed in Sections 1.1 and 1.2. Sections 1.3 and 1.4 describe the analysis and design of transversal filters. It may be useful to review this chapter before Chapter 9 on bandpass filters.

1.1. Frequency Response and Impulse Response

Filters are linear devices; an input signal at a single frequency produces an output signal at the same frequency, but with a different amplitude and phase. The change in amplitude and phase depends on the frequency of the signal and can be described by a complex function of the frequency f, $H(f)$, which is called the *frequency response* of the filter. Filters are usually specified in terms of their frequency response. An alternative way of specifying a filter is in terms of its *impulse response*, which describes the "memory" of the filter; it tells us how the output at the present time is affected by an input at a past time. Any signal processing device that discriminates between signals of different frequencies must have memory. At a single instant in time, signals of all frequencies look alike; to tell the difference between two frequencies f_1 and f_2 the signal has to be watched, at least for a length of time of order $|f_1 - f_2|^{-1}$. A broadband amplifier, by contrast, can operate on an instant-by-instant basis, with no memory.

A conventional RLC filter (Fig. 1.1) derives its memory from a resonant circuit. If the input signal $v_i(t)$ is an impulse (Fig. 1.1b), the output $v_0(t)$ rings for a while before it decays (Fig. 1.1c). This is called the impulse response, $h(t)$, of the filter. The output has a memory of its past inputs and the impulse response is a way of describing this memory. This circuit acts as a bandpass filter, passing frequencies close to the frequency of the ringing. If the resistance R is very small, the circuit rings for a long time (has a long memory) and the filter is very narrowband; it can discriminate between frequencies very close together. If the resistance R is large, the ringing dies out quickly and the filter is wideband.

A filter can be completely characterized by either its frequency response, $H(f)$, or its impulse response, $h(t)$; they are Fourier transforms of each other (see Example 1.2).

$$h(t) = \int_{-\infty}^{+\infty} df \, H(f) e^{+j2\pi ft} \qquad (1.1a)$$

$$H(f) = \int_{-\infty}^{+\infty} dt \, h(t) e^{-j2\pi ft} \qquad (1.1b)$$

A well-known property of the Fourier transform is that the longer $h(t)$ is in the time domain, the narrower $H(f)$ is in the frequency domain. This leads to the property we have discussed earlier; that a filter with a longer memory is more narrowband in its frequency response.

Example 1.1
A filter is excited with an input signal at 100 MHz described by

$$v_i(t) = \cos(2\pi \times 10^8 t)$$

1 TRANSVERSAL FILTERS

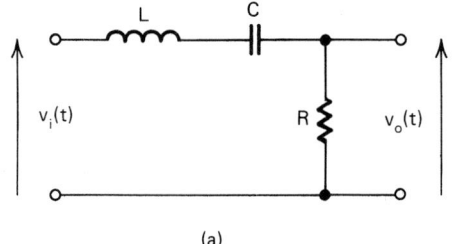

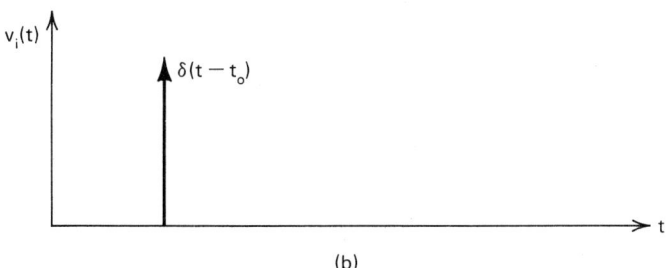

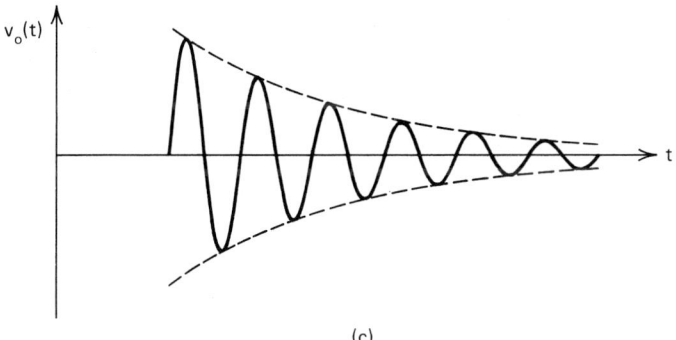

Figure 1.1 RLC filter: (a) circuit diagram; (b) impulse input at $T = T_o$; (c) impulse response.

The output signal is measured to be (Fig. 1.2)

$$v_0(t) = \frac{1}{2} \cos\left(2\pi \times 10^8 t - \frac{\pi}{6}\right)$$

Calculate $H(f)$ at $f = 100$ MHz.

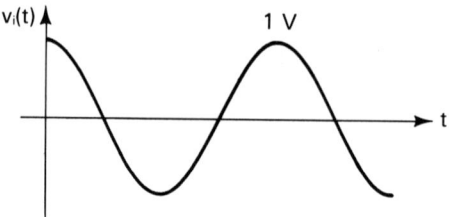

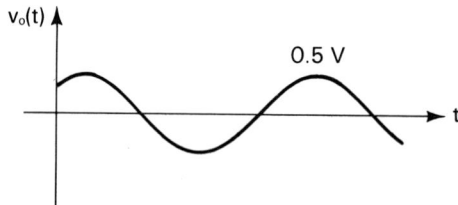

Figure 1.2 Input and output signals in Example 1.1.

Solution

By inspection we can see that the output has half the amplitude of the input and is shifted by $\dfrac{\pi}{6}$ in phase. Hence

$$H(f = 100 \text{ MHz}) = \frac{1}{2} e^{-j\pi/6}$$

It is usual to represent real-time signals in terms of complex exponential signals. For example, we can write $v_i(t)$ and $v_o(t)$ as

$$v_i(t) = \frac{1}{2} e^{j2\pi \times 10^8 t} + \frac{1}{2} e^{-j2\pi \times 10^8 t}$$

$$v_o(t) = \left[\frac{1}{4} e^{-j\pi/6}\right] e^{j2\pi \times 10^8 t} + \left[\frac{1}{4} e^{j\pi/6}\right] e^{-j2\pi \times 10^8 t}$$

In general, we can write a signal $v(t)$ at a frequency f as

$$v(t) = V(f) e^{j2\pi f t} + V(-f) e^{-j2\pi f t}$$

where *V(-f)* is equal to $V^*(f)$, the complex conjugate of *V(f)*; this is necessary in order that $v(t)$ comes out real, as all actual time signals must.

1 TRANSVERSAL FILTERS

Thus a real-time signal at a frequency f can be considered as the sum of a complex exponential signal at $+f$ and another at $-f$, the coefficients of the two being complex conjugates of each other. In our present example

$$V_i(f) = \frac{1}{2} \qquad V_i(-f) = V_i^*(f) = \frac{1}{2}$$

$$V_0(f) = \frac{1}{4} e^{-j\pi/6} \qquad V_0(-f) = V_0^*(f) = \frac{1}{4} e^{j\pi/6}$$

The frequency response $H(f)$ is defined as

$$H(f) = \frac{V_0(f)}{V_i(f)} \qquad (1.2)$$

$$H(-f) = \frac{V_0(-f)}{V_i(-f)} = H^*(f)$$

In the present example this yields $H(f) = \frac{1}{2} e^{-j\pi/6}$, in agreement with what we wrote down intuitively earlier.

This concept can be extended to nonsinusoidal time signals. In general, any signal $v(t)$ can be written as a superposition of sinusoidal signals of different frequencies. Mathematically, this is represented as an integral over a range of frequencies.

$$v(t) = \int_{-\infty}^{+\infty} V(f) e^{j2\pi ft} \, df \qquad (1.3a)$$

Once again to ensure that $v(t)$ is real we must have $V(-f) = V^*(f)$. Equation 1.3a shows that $v(t)$ is the inverse Fourier transform of $V(f)$, which means that $V(f)$ must be the Fourier transform of $v(t)$:

$$V(f) = \int_{-\infty}^{+\infty} v(t) e^{-j2\pi ft} \, dt \qquad (1.3b)$$

Example 1.2
Prove Eq. 1.1.

Solution
Consider a filter with frequency response $H(f)$. We wish to show that the output $v_0(t)$ in response to an impulse input $v_i(t) = \delta(t)$ is $h(t)$, the Fourier transform of $H(f)$.

$$v_i(t) = \delta(t)$$

$$V_i(f) = \int_{-\infty}^{+\infty} \delta(t)e^{-j2\pi ft}\, dt \qquad (1.3b)$$

$$= 1$$

$$V_o(f) = V_i(f)H(f) = H(f) \qquad (1.2)$$

$$v_o(t) = \int_{-\infty}^{+\infty} H(f)e^{j2\pi ft}\, df \qquad (1.3a)$$

$$= h(t) \qquad (1.1a)$$

Example 1.3
Suppose that a filter has the impulse response $h(t)$ shown in Fig. 1.3.

$$h(t) = \begin{cases} \alpha e^{-\alpha t} & t \geq 0 \\ 0 & t < 0 \end{cases}$$

Calculate the frequency response $H(f)$.

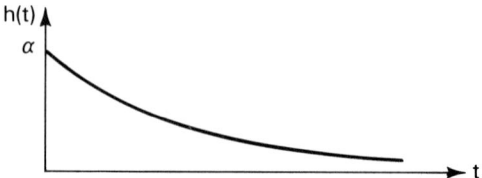

Figure 1.3 Impulse response in Example 1.3.

Solution
From Eq. 1.1b,

$$H(f) = \frac{\alpha}{\alpha + j2\pi f}$$

so that the magnitude $|H(f)|$ and the phase $\underline{/H(f)}$ are written as

$$|H(f)| = \frac{\alpha}{(\alpha^2 + 4\pi^2 f^2)^{1/2}}$$

$$\underline{/H(f)} = \arctan\left[-\frac{2\pi f}{\alpha}\right]$$

1 TRANSVERSAL FILTERS

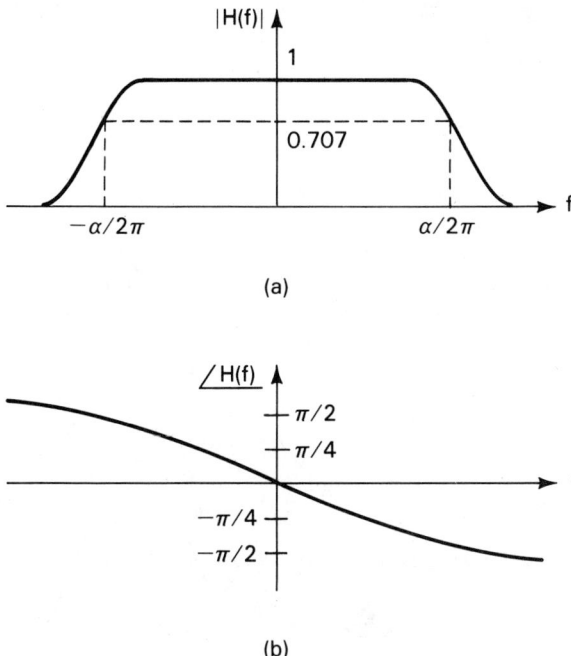

Figure 1.4 Frequency response of a filter with the impulse shown in Fig. 1.3: (a) magnitude; (b) phase.

These are shown in Fig. 1.4. Note that the phase does not change linearly with frequency. In general, it can be shown that if the impulse response is symmetric about any point in time, the frequency response has a linear phase. But if there is no point of symmetry (as in this example), the phase is nonlinear. The response considered in this example represents a typical RC low-pass filter, α^{-1} being the time constant RC.

1.2. Some Properties of the Fourier Transform

We have seen that the frequency response $H(f)$ is the Fourier transform of the impulse response $h(t)$. There are certain general properties that hold for all Fourier transform pairs which are very useful in visualizing how certain changes in $h(t)$ affect $H(f)$, and vice versa. We will now discuss some of these properties.

1.2.1. Shifting Theorem

If we shift $h(t)$ in time by t_0, the magnitude of $H(f)$ is unaffected but its phase is changed by $(-2\pi f t_0)$. For example, suppose that the impulse response in Example 1.3 does not start at $t = 0$, but starts at $t = t_0$. The frequency response then becomes

$$H(f) = \frac{\alpha}{\alpha + j2\pi f} e^{-j2\pi f t_0}$$

so that

$$|H(f)| = \frac{\alpha}{(\alpha^2 + 4\pi^2 f^2)^{1/2}} \quad \text{(same as before)}$$

$$\underline{/H(f)} = -2\pi f t_0 + \arctan\left[-\frac{2\pi f}{\alpha}\right]$$

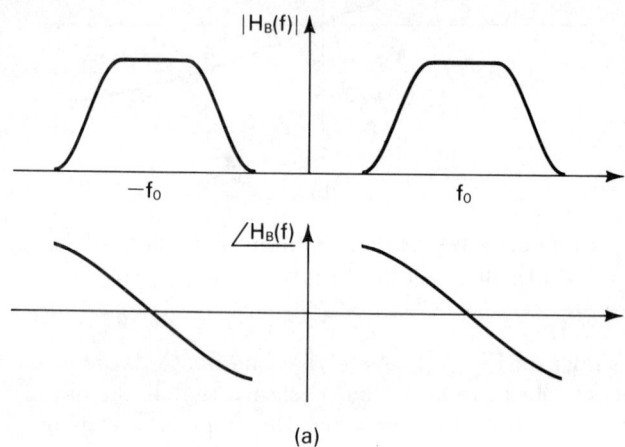

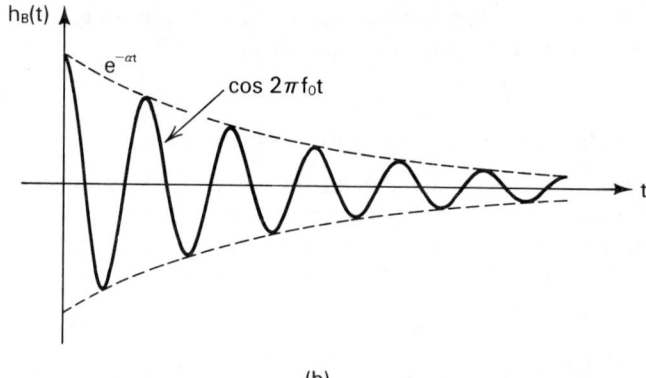

Figure 1.5 Bandpass filter: (a) frequency response; (b) impulse response.

1 TRANSVERSAL FILTERS

A similar result also holds for shifting in frequency. If $H(f)$ is shifted by f_0, the impulse response, $h(t)$, is multiplied by $e^{+j2\pi f_0 t}$. This needs a little explanation since we know that the impulse response must be real. In Example 1.3 we considered a low-pass filter. Suppose that we wish to find the impulse response $h_B(t)$ of a bandpass filter with a frequency response $H_B(f)$ which is exactly the same as $H(f)$ but is centered around $f = f_0$ rather than around $f = 0$. However, we have to remember that $H_B(-f)$ must be equal to $H_B^*(f)$, so that we also need a response centered around $f = -f_0$ (Fig. 1.5).

$$H_B(f) = H(f - f_0) + H(f + f_0)$$

Noting that a shift in $H(f)$ by f_0 means a multiplication of $h(t)$ by $e^{+j2\pi f_0 t}$, we have

$$h_B(t) = h(t)e^{j2\pi f_0 t} + h(t)e^{-j2\pi f_0 t}$$

$$= 2h(t)\cos 2\pi f_0 t$$

The impulse response of the bandpass filter is shown in Fig. 1.5b. This is the kind of response we get from an RLC filter (Fig. 1.1).

1.2.2. Convolution Theorem

Suppose that we cascade two filters. The overall frequency response $H(f)$ is the product of the individual responses $H_1(f)$ and $H_2(f)$.

$$H(f) = H_1(f) \cdot H_2(f)$$

We would like to know how the overall impulse response $h(t)$ is related to the individual responses $h_1(t)$ and $h_2(t)$. The convolution theorem states that

$$h(t) = h_1(t) * h_2(t)$$

where $*$ denotes the convolution product, which is defined as

$$h_1(t) * h_2(t) = \int_{-\infty}^{+\infty} d\tau\, h_1(\tau) h_2(t - \tau) \tag{1.4}$$

The property (like most properties of the Fourier transform) also holds in reverse; that is, if

$$h(t) = h_1(t) \cdot h_2(t)$$

then

$$H(f) = H_1(f) * H_2(f)$$

We can use this property to write the output $v_o(t)$ of a filter, with impulse response $h(t)$, to an arbitrary input $v_i(t)$. We have seen that (Eq. 1.2)

$$V_o(f) = H(f)V_i(f)$$

From the convolution theorem,

$$\begin{aligned} v_o(t) &= v_i(t) * h(t) \\ &= \int_{-\infty}^{+\infty} d\tau \, v_i(t) h(t-\tau) \end{aligned} \quad (1.5)$$

This has an interesting interpretation. We know that $h(t)$ represents the memory of the filter; $h(t-\tau)$ tells us the output at time t due to a unit impulse input at time τ. The output at time t due to an input $v_i(\tau)$ at time τ is thus given by $v_i(\tau) \cdot h(t-\tau)$. To get the total output, we have to sum the outputs resulting from all the past inputs; that is, we have to integrate $v_i(\tau)h(t-\tau)$ over all τ.

Example 1.4 Graphical Convolution
Consider the low-pass filter in Example 1.3. Find the output for the input shown in Fig. 1.6a.

Solution
The integration in Eq. 1.5 can be carried out either graphically or analytically. Analytically, we have

$$h(t-\tau) = \begin{cases} \alpha e^{-\alpha(t-\tau)} & \tau \leq t \\ 0 & \tau > t \end{cases}$$

$$v_i(\tau) = \begin{cases} 1 & 0 \leq \tau \leq T \\ 0 & \tau < 0, \tau > T \end{cases}$$

Hence

1 TRANSVERSAL FILTERS

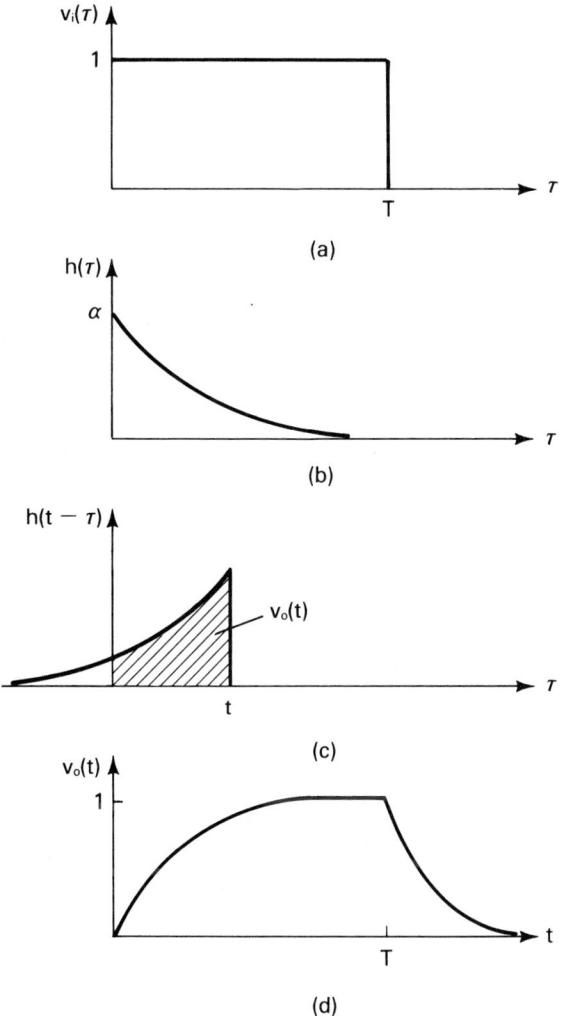

Figure 1.6 Graphical convolution: (a) input signal; (b) impulse response; (c) time-reversed and shifted impulse response; (d) output signal.

$$v_o(t) = \begin{cases} \int_0^t d\tau\, \alpha e^{-\alpha(t-\tau)} & 0 < t \leq T \\ \int_0^T d\tau\, \alpha e^{-\alpha(t-\tau)} & t \geq T \end{cases}$$

That is,

$$v_o(t) = \begin{cases} 1 - e^{-\alpha t} & 0 < t \leq T \\ (1 - e^{-\alpha T})e^{-\alpha(t - T)} & t \geq T \end{cases}$$

The output, $v_o(t)$, is shown in Fig. 1.6d assuming that $\alpha T \gg 1$, so that $e^{-\alpha T}$ is negligible. The waveform will be recognized as the familiar rise and decay curves of an RC circuit. The graphical procedure is shown in Fig. 1.6. We can time reverse the $h(t)$ response [because the argument of $h(t - \tau)$ decreases with increasing τ] and slide it over $v_i(\tau)$ (to correspond to increasing t). The value of $v_o(t)$ corresponds to the area of the curve obtained from the product of these two curves. This can be best illustrated by sliding a (flipped) transparency of $h(t)$ over $v_i(t)$, taking the product of the two and using the area under that curve.

1.2.3. Sampling Theorem

Let us consider a filter whose impulse response looks similar to that in Example 1.3 but is discrete rather than continuous (Fig. 1.7a). The sampling theorem tells us that the corresponding frequency response is periodic (Fig. 1.7b) with the same passband repeated at frequency intervals of $1/T$, where T is the spacing between the time samples. Mathematically,

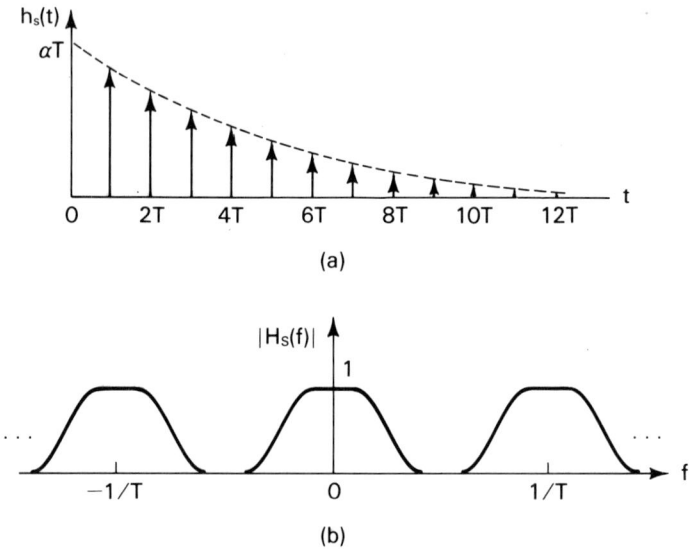

Figure 1.7 (a) Discrete impulse response; (b) corresponding periodic frequency response.

1 TRANSVERSAL FILTERS

the frequency response $H_S(f)$ corresponding to the sampled impulse response is given by

$$H_S(f) = \sum_{n=-\infty}^{+\infty} H(f - nf_s) \tag{1.6}$$

where $H(f)$ is the frequency response corresponding to the continuous impulse response and f_s is the sampling frequency $(= 1/T)$. Provided that the sampling frequency f_s is much greater than the filter bandwidth α/π, the individual band shapes faithfully replicate the original passband; but if the samples are too far apart (making f_s too small), there is considerable overlap between the different passbands, causing what is known as the *aliasing error*.

Example 1.5
Consider the impulse response in Fig. 1.8. Calculate the corresponding frequency response.

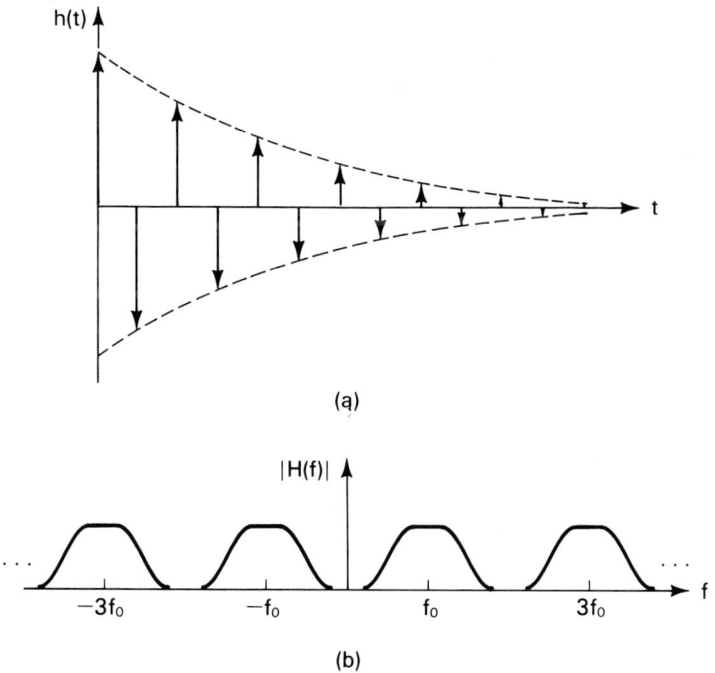

Figure 1.8 (a) Discrete impulse response; (b) corresponding periodic frequency response.

Although the subscript is not italicized, it is a variable.

Solution
Note that the impulse response is a sampled version of the impulse response of the bandpass filter (Fig. 1.5) discussed earlier; the sampling frequency is

$$f_s = 2f_0$$

since there are two samples (one positive and one negative) in every cycle. Hence the frequency response will be periodic with a period of $2f_0$; passbands appear around all odd multiples of f_0.

1.3. Recursive and Nonrecursive Filters

We will now discuss a different approach to making filters, using delay lines rather than resonant circuits to provide the memory. Consider the simple setup in Fig. 1.9a using a feedback loop containing a delay line with delay time $= T$ and an attenuator with attenuation $= e^{-\alpha T}$. Let us try to figure out its impulse response. If the input is an impulse at $t = 0$, it immediately appears at the output. The signal then goes around the feedback loop and appears again at the output after time T, reduced in amplitude by $e^{-\alpha T}$. This continues on giving rise to the impulse response shown in Fig. 1.9b.

Note that this impulse response is a sampled version of that of an RC filter with time constant α^{-1}. We discussed the corresponding frequency response in Section 1.2. The response is periodic in frequency with

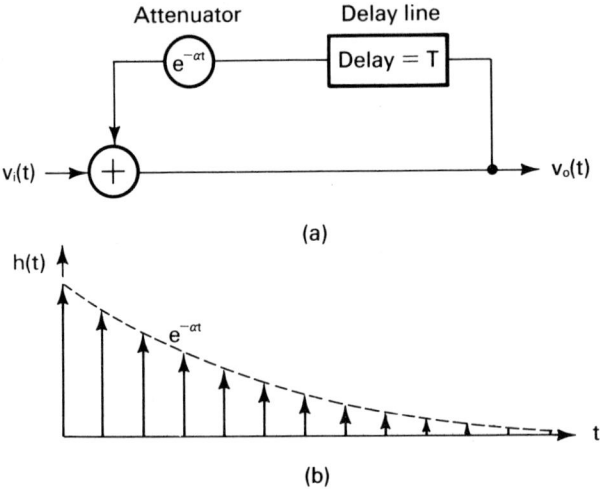

Figure 1.9 Simple digital filter: (a) configuration; (b) impulse response.

1 TRANSVERSAL FILTERS

passbands around $f = nf_s$, where f_s is the sampling frequency ($= 1/T$) and n is an integer (Fig. 1.7). By putting a broadband filter at the input, we can select any of these passbands and use it as either a low-pass filter or a bandpass filter.

Example 1.6
Suppose that the attenuator in the feedback loop is $-e^{-\alpha T}$ rather than $e^{-\alpha T}$ in Fig. 1.9a. How does this affect the frequency response?

Solution
In this case the samples alternate in sign, giving the impulse response in Fig. 1.8a rather than the one in Fig. 1.7a. The frequency response will thus be the one in Fig. 1.8b rather than the one in Fig. 1.7b.

What we have just discussed is called a *recursive* digital filter. We also have nonrecursive digital filters, in which there is no feedback from the output to the input. The output is formed by summing weighted samples of the input taken at various points along a delay line (Fig. 1.10a). This gives the impulse response shown in Fig. 1.10b. Here we have shown only a seven-tap delay line; by increasing the number of taps, more complicated impulse response shapes can be synthesized. Such a filter is also called a *transversal* filter and was first proposed by Kallman in 1940 (Ref. 1.1). Its utility, however, was limited since the total delay-time variable in electromagnetic delay lines with practical physical dimensions is rather small because of the large propagation velocity of electromagnetic waves.

This difficulty is overcome by the use of acoustic delay lines. The electrical signal is first converted to an acoustic wave; this is done conveniently in a class of solids called *piezoelectric solids* in which an electric field produces a mechanical stress, and vice versa. Acoustic waves are five orders of magnitude slower than electromagnetic waves, so that large delays are obtained with compact practical devices. Moreover, the *surface acoustic wave* (SAW) propagates along the surface so that the signal is accessible at every point along the propagation path. An interdigitated pattern of metallic (aluminum, gold, etc.) electrodes is fabricated on the surface of a piezoelectric solid (lithium niobate, quartz, etc.) using photolithographic techniques; these act as transducers that convert electrical signals to acoustic waves, and vice versa (Fig. 1.10c). They are called *interdigital transducers* (IDTs) and are used to tap the SAW energy along the propagation path; the summation is performed automatically by the busbars connecting the electrodes. The samples can be weighted by varying the length of the electrodes, thus providing a compact realization of the transversal filter concept. SAW devices have developed rapidly since their conception in the mid-1960s and are widely used in making bandpass filters, resonators, oscillators, and matched filters. They are chiefly used in the VHF/UHF range, where the acoustic

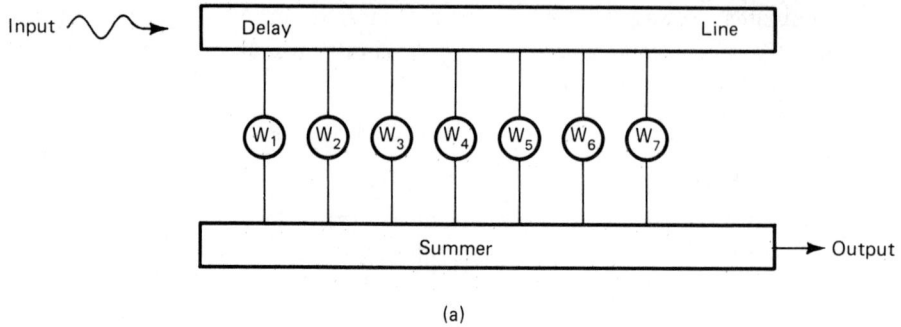

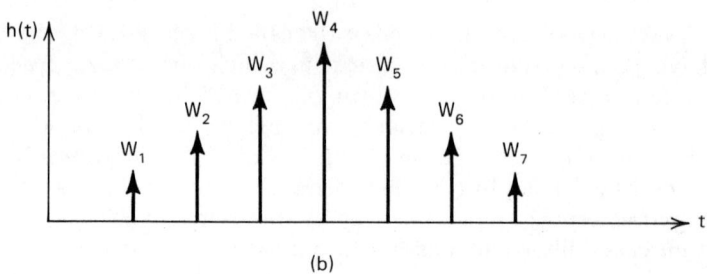

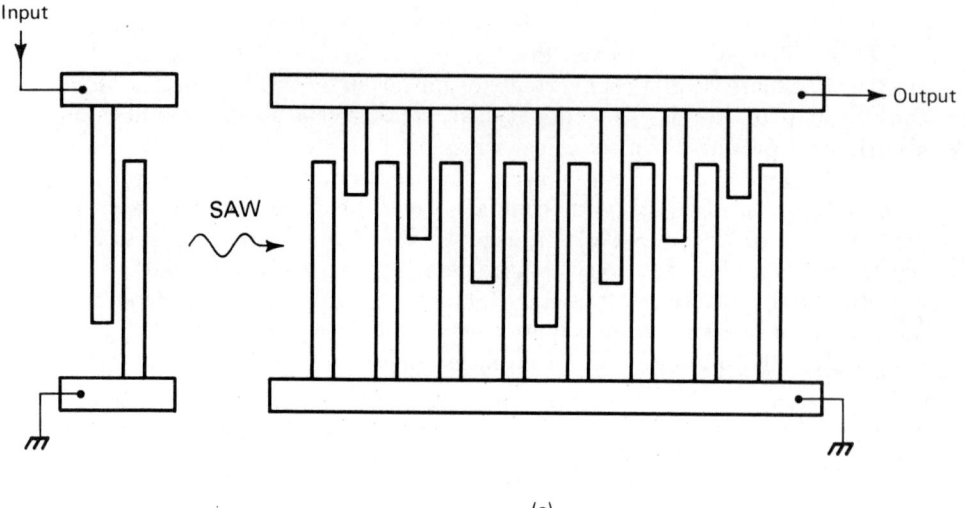

Figure 1.10 (a) Transversal filter; (b) impulse response; (c) SAW delay line implementation of the transversal filter.

1 TRANSVERSAL FILTERS

wavelength is small enough to make device compact and yet not so small as to make photolithography difficult.

One advantage of transversal filters is that arbitrary impulse response shapes can be synthesized in a straightforward manner. The impulse response can be made symmetric about the center by choosing the tap weights appropriately (for example, by making $W_5 = W_3$, $W_6 = W_2$, and $W_7 = W_1$ in Fig. 1.10). A symmetric impulse response $h(t)$ leads to a frequency response $H(f)$ whose phase varies linearly with frequency. In fact,

$$\underline{/H(f)} = -2\pi f t_0$$

if the center of the impulse response occurs at $t = t_0$. This linear phase characteristic is extremely important in many filtering applications and is difficult to achieve using either conventional RLC filters or recursive digital filters.

Example 1.7
Consider a transversal filter with $(N+1)$ taps all of equal weight and separated by delays of T. The impulse response $h_s(t)$ is shown in Fig. 1.11a. Calculate the frequency response.

Solution
Let us first consider the continuous version of $h_s(t)$, which we will call $h(t)$ (Fig. 1.11b). After all, sampling at intervals of T merely means that the passband is replicated at frequency intervals of $f_s = 1/T$. The frequency response is obtained using Eq. 1.3b.

$$H(f) = \int_0^{NT} dt \, e^{-j2\pi f t}$$

$$= \frac{N}{f_s} e^{-jN\pi f/f_s} \frac{\sin(N\pi f/f_s)}{N\pi f/f_s}$$

where $f_s = 1/T$. The magnitude of $H(f)$ is shown in Fig. 1.11c. The magnitude is plotted in decibels, showing that the first sidelobe is 13 dB below the center of the passband. The phase changes linearly with frequency (since the impulse response is symmetric about $t = NT/2$) with abrupt phase reversals at the points where the magnitude goes to zero (f_s/N, $2f_s/N$, etc.). The width of the response from the center to the first null is f_s/N; this also happens to be the 4-dB bandwidth of the filter. The 3-dB bandwidth BW_{3dB} is a little smaller:

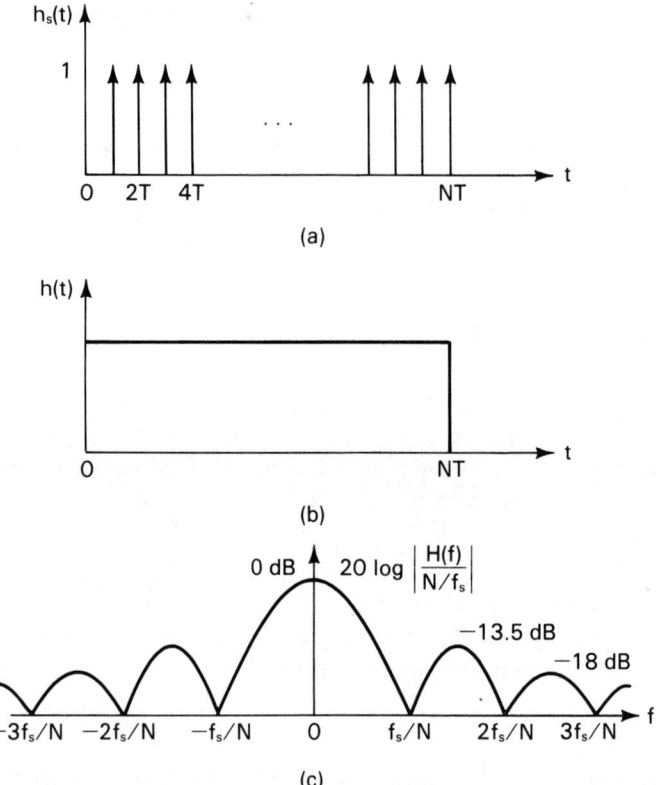

Figure 1.11 Equally weighted N-tap transversal filter: (a) impulse response $H_s(t)$ of transversal filter; (b) corresponding continuous impulse response $h(t)$; (c) frequency response.

$$\text{BW}_{3\text{dB}} = 0.9 \frac{f_s}{N}$$

The actual frequency $H_S(f)$ corresponding to the sampled impulse response $h_s(t)$ has multiple passbands centered around $f = \pm n f_s$, n being an integer. Provided that N is large, so that the filter bandwidth is much smaller than the sampling frequency, there is no significant aliasing effect.

Example 1.8
We wish to design an interdigital transducer (IDT) for a SAW delay line (Fig. 1.12) so that it has a $(\sin x)/x$ frequency response centered around 100 MHz and a 4-dB bandwidth of 10 MHz. Calculate the number of

1 TRANSVERSAL FILTERS

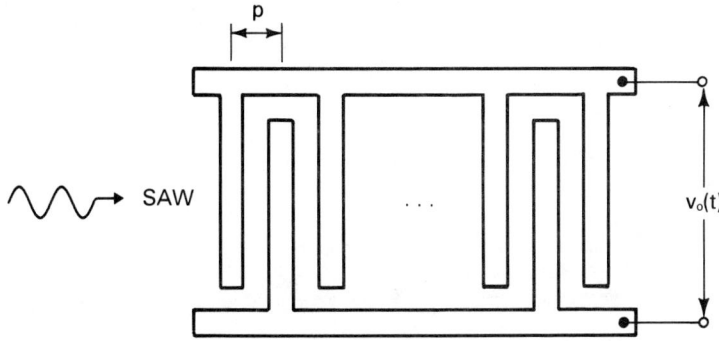

Figure 1.12 SAW delay-line implementation of the transversal filter in Example 1.7.

electrodes and the spacing between the electrodes in the IDT. SAW velocity $v_0 = 4000$ m/s.

Solution
As we know, passbands are obtained around frequencies $f = \pm n f_s$; so to get a response centered at 100 MHz, we could use any $f_s = 100$ MHz/n. However, the conclusions of Example 1.7 are based on the assumption that each sample is infinitely narrow in time, which is not true for SAW devices, as we will see in Chapter 4. This affects the passbands at higher multiples of f_s. For the present let us assume that $n = 1$, so that $f_s = 100$ MHz. This means that the delay T from one positive electrode to the next is

$$T = \frac{1}{f_s} = 0.01 \ \mu s$$

Hence the spacing between electrodes p is given by

$$p = \frac{T}{2} v_0 = 20 \ \mu m$$

We have divided by 2 because T corresponds to the delay between positive electrodes. The 4-dB bandwidth is equal to f_s/N. Hence $N = 10$, which means that we need 10 positive electrodes in all. Usually, IDTs are made with a 50% metallization ratio, which means that the electrodes and gaps are of equal width. So in this example, the electrodes are each of width 10 μm.

Example 1.9
Assuming that the smallest electrode width that can be fabricated with photolithographic techniques is 1 μm, what is the highest frequency at which SAW filters can be built operating at the fundamental frequency?

Solution
We saw in Example 1.8 that a 100-MHz filter needs an electrode width of 10 μm. Since distances scale inversely with frequency, 1-μm electrode widths will give a 1-GHz filter.

1.4. Design of Nonrecursive (Transveral) Filters

It will be noted that unlike RLC filters or recursive filters, the impulse response of transversal filters is always finite in length. For this reason these are also known as *finite impulse response* (FIR) filters. If we take an arbitrary frequency response, the corresponding impulse response will usually extend to infinity. To implement it as a transversal filter, it is necessary to truncate the impulse response. The obvious way to do this is to cut it off at a point where its value has grown sufficiently small. However, such an abrupt truncation of the impulse response produces large ripples in the frequency response, and window functions are often used to provide a gradual cutoff.

Suppose that we wish to build an ideal brickwall filter (Fig. 1.13a) centered at $f = 0$. In practice, we are usually interested in bandpass filters, but this merely involves a shift in frequency.

$$H(f) = \begin{cases} 1 & -\Delta f/2 < f < +\Delta f/2 \\ 0 & \text{otherwise} \end{cases}$$

The corresponding impulse response is given by (Fig. 1.13a)

$$h(t) = \Delta f \frac{\sin \pi \Delta f t}{\pi \Delta f t}$$

Incidentally, the reason we have an $h(t)$ symmetric around $t = 0$ is that we neglected the phase of $H(f)$. If we assumed that $\underline{/H(f)} = -2\pi f t_0$, we would get the same impulse response, but centered around $t = t_0$.

The impulse response extends from negative to positive infinity and we need to truncate it before we can implement it as a transversal filter. We wish to find the best way to truncate it so as to cause minimal distortion in the frequency response, and this involves the use of window functions. Let us denote the window function by $w(t)$, so that the truncated impulse response $h_T(t)$ is given by

1 TRANSVERSAL FILTERS

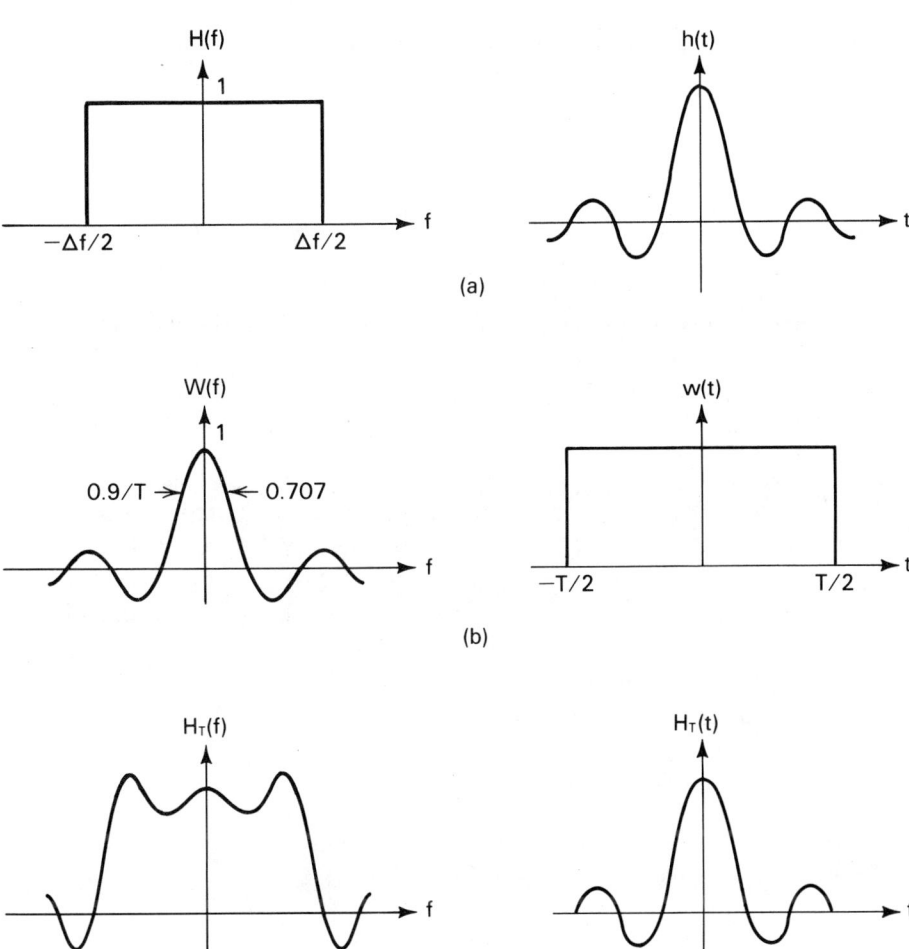

Figure 1.13 (a) Brickwall response; (b) rectangular window function; (c) truncated response.

$$h_T(t) = h(t) \cdot W(t)$$

The corresponding frequency response $H_T(f)$ is given by the convolution theorem.

$$H_T(f) = H(f) * W(f)$$

$$= \int_{-\infty}^{+\infty} df' \; W(f')H(f-f')$$

where $W(f)$ is the Fourier transform of the window function. Since $H(f)$ is zero outside the range $-\Delta f/2 < f < +\Delta f/2$ and is unity inside the range, we have

$$H_T(f) = \int_{f-\Delta f/2}^{f+\Delta f/2} df' \, W(f')$$

This means that the actual frequency response $H_T(f)$ after truncation is equal to the average value of the Fourier transform of the window function in the range $f \pm \Delta f/2$. Clearly, then, the frequency response $H_T(f)$ cannot change any more abruptly or be any flatter than $W(f)$.

For example, Fig. 1.13 shows the result of abrupt truncation using a rectangular window for $W(t)$. The Fourier transform $W(f)$, as we have seen before, is a sinc function with the first sidelobe 13 dB below the main lobe. The 3-dB bandwidth of the main lobe is $0.9/T$ where T is the length of the gate in time (Example 1.7). When this $W(f)$ is convolved with $H(f)$ we get a frequency response $H_T(f)$ (Fig. 1.13c) with a transition bandwidth BW $_T$ equal to the width of the main lobe of $W(f)$; that is, the sidelobes of $H_T(f)$ are as big as those of $W(f)$, but 13 dB below the main response. The sidelobes of $W(f)$ also cause ripples in the passband of $H_T(f)$; -13-dB sidelobes will produce a 1.7-dB peak ripple (see Example 1.11). Making the window longer does not help much either; it reduces BW $_T$ ($= 0.9/T$), but we still have -13-dB sidelobes and a 1.7-dB peak ripple.

To reduce the ripple and the sidelobe levels, different window functions $W(t)$ are used whose Fourier transforms $W(f)$ have low sidelobe levels. For example, the Hamming window is very popular, with sidelobes -42.8 dB below the passband. It is given by (Fig. 1.14)

$$w(t) = 0.54 + 0.46 \cos \frac{2\pi t}{T}$$

It will be noted that the main lobe (in frequency) is 1.47 times wider than that of the rectangular window (Fig. 1.13b) for the same window length T. For the same T, we thus get a longer transition bandwidth using a Hamming window than using a rectangular window. However, we can

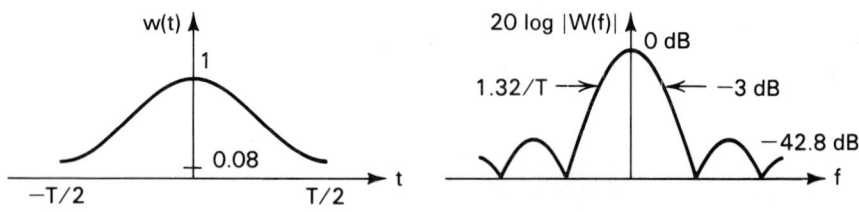

Figure 1.14 Hamming window.

1 TRANSVERSAL FILTERS

always make the window long enough to get the desired transition bandwidth.

Using different window functions, it is also possible to trade off transition bandwidth for sidelobe level with a fixed window length T. For example, the coefficients 0.54 and 0.46 in the Hamming window are the ones that yield the lowest sidelobe levels for this class of functions; using a different set of coefficients we could get smaller transition bandwidths but higher sidelobe levels. A good rule of thumb is that in order to get a frequency response $H_T(f)$ with a ripple magnitude R and a transition bandwidth BW_T (Fig. 1.15), we need a window function that has a length of at least

$$T_{min} = \frac{0.8}{BW_T} \left[\frac{\log 2}{R} \right]^{1/2} \tag{1.7}$$

If the ripple magnitude is different in the passband than outside, the smaller one should be used.

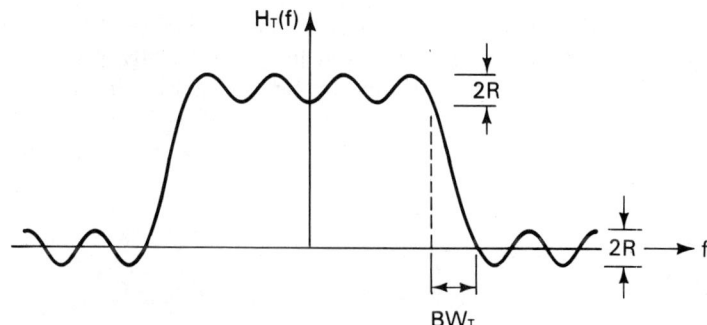

Figure 1.15 Filter response showing ripple and transition bandwidth.

Equation 1.7 gives us the shortest impulse response length that could be obtained using an optimal window function. This optimal function is called the Dolph–Chebyshev window and is physically unrealizable because it is infinitely large at the two edges. It is usually used as a standard against which the optimality of other window functions can be tested.

Example 1.10
Check the optimality of the rectangular and Hamming windows.

Solution
For the rectangular window,

$$\text{sidelobe level} = 20 \log \frac{1}{R} = 13 \text{ dB}$$

$$BW_T = 0.9/T$$

$$T_{min} = 0.87\ T \qquad (1.7)$$

In other words, an optimal window would give the same sidelobe and transition bandwidth with an impulse response 87% as long. For the Hamming window,

$$20 \log \frac{1}{R} = 42.8$$

$$BW_T = \frac{1.32}{T}$$

$$T_{min} = 0.95\,T$$

Example 1.11
We wish to design a filter with a 1-MHz transition bandwidth, sidelobes at least 40 dB below the main response, and less than 0.2 dB of peak ripple in the passband. Calculate the minimum length of the impulse response required.

Solution

$$BW_T = 1\ \text{MHz}$$

From the sidelobe specification we have the ripple R_{SL} in the sidelobes.

$$20 \log_{10} R_{SL} = -40$$

$$R_{SL} = 0.01$$

From the passband specification, we get the ripple R_p in the passband.

$$20 \log_{10}(1 + R_p) = 0.2$$

Since R_p is larger than R_{SL}, it is the sidelobe specification that is more stringent and we use

$$R = R_{SL}\ (0.01)$$

1 TRANSVERSAL FILTERS

Using Eq. 1.7 yields

$$T_{min} = 1.2 \; \mu s$$

Example 1.12
If a SAW device is used to implement the impulse response in Example 1.11, how long would the IDT be? (The SAW velocity $v_0 = 4000$ m/s.)

Solution
To get an impulse 1.2 μs long, the length L of the IDT will be

$$L = 1.2 \; \mu s \times 4000 \; m/s$$

$$= 0.48 \; cm$$

So the SAW device would be at least $\frac{1}{2}$ cm long. In fact, it will be somewhat longer since there is some spacing between the input and output transducers; what we just calculated was the length of the output transducer, assuming a short-broadband input transducer.

Example 1.13
Assuming that the longest available substrate is 2 in. long, what is the minimum transition bandwidth that we can get from a SAW device with the same ripple and sidelobe specifications as those in Example 1.11?

Solution
The transition bandwidth varies inversely as the length. Since a BW $_T$ of 1 MHz requires $\frac{1}{2}$ cm, 2 in. will give

$$BW_T = 100 \; Hz$$

This illustrates the low-frequency limit to SAW devices, since low-frequency filters also have small transition bandwidths (absolute, not relative) and this requires too long a crystal.

Window functions provide a convenient design method for FIR filters. However, it has been pointed out that the windowing technique produces the largest error in the frequency response $H_T(f)$ around the transition regions; a better filter (with lower maximum error) could be designed using an optimization algorithm that spreads out the error uniformly over frequency. A widely used algorithm is one developed by Remez and extended by McClellan et al. (Ref. 1.2).

It is apparent from the preceding discussion that the analysis and design of transversal filters requires frequent transformations between time and frequency domains. This is usually done on a digital computer using fast Fourier transform (FFT) algorithms that require the information to be sampled in both domains. The following examples illustrate the relevant considerations in choosing the number of samples and sampling rates.

Example 1.14
We have a frequency-domain response of a bandpass filter with a center frequency of 100 MHz and a transition bandwidth of 1 MHz. We wish to transform it to time domain to determine the tap weights needed to implement it as a SAW filter. Calculate the number of sample points required if the SAW filter has two taps per cycle.

Solution

Let F = total range in frequency domain

Δf = spacing between samples in frequency domain

T = total range in time domain

Δt = spacing between samples in time domain

From the sampling theorem (Section 1.2.3) it is clear that $T = (\Delta f)^{-1}$ and $F = (\Delta t)^{-1}$, so that

$$N = \frac{F}{\Delta f} = \frac{T}{\Delta t} = FT = \frac{1}{\Delta f \, \Delta t}$$

where N is the number of sample points in both the time and frequency domains. We can determine N starting from either the time or the frequency domain. Let us start with the time domain. The length of the time response T is approximately 1.2 μs (Example 1.11). The sampling rate Δt is equal to $(2f_c)^{-1} = 0.005$ μs, since there are two taps per cycle and the center frequency $f_o = 100$ MHz. Hence $N = 240$. FFT algorithms require the number of points to be an integer power of 2 so that an FFT with $N = 2^8$ can be used. In the frequency domain the sample spacing Δf is equal to $1/T$ ($\simeq 0.8$ MHz), which is of the order of the transition bandwidth, while the frequency range is equal to $1/\Delta t$ ($\simeq 200$ MHz). Note that the frequency range F is two times the

1 TRANSVERSAL FILTERS

center frequency, which is much larger than the bandwidth. This is necessary to ensure a one-for-one correspondence between the time-domain samples and the electrodes of the SAW filter. However, in principle, we should be able to reduce F to the order of the bandwidth with no loss of information; in the time domain this means that the samples vary slowly in magnitude and that one can sample less frequently with no loss of information.

In practice, one should use a bigger time range (T) and a closer frequency sampling (Δf), to ensure that the impulse response is indeed negligible outside this range; otherwise, any response outside of T will be aliased back into our range of interest. This is particularly true if the frequency response is an experimentally measured one having spurious echoes occurring at long times.

Example 1.15
Figure 1.16b shows the Fourier transform of the frequency response in Fig. 1.16a. What is the frequency range used in the FFT?

Solution
It is apparent from the time response that there are two samples per period. In a SAW filter these correspond to positive and negative electrodes. The sampling rate $\Delta t = 0.01$ μs, since the center frequency $f_c = 50$ MHz. Hence, the frequency range F must be equal to $(\Delta t)^{-1} = 100$ MHz.

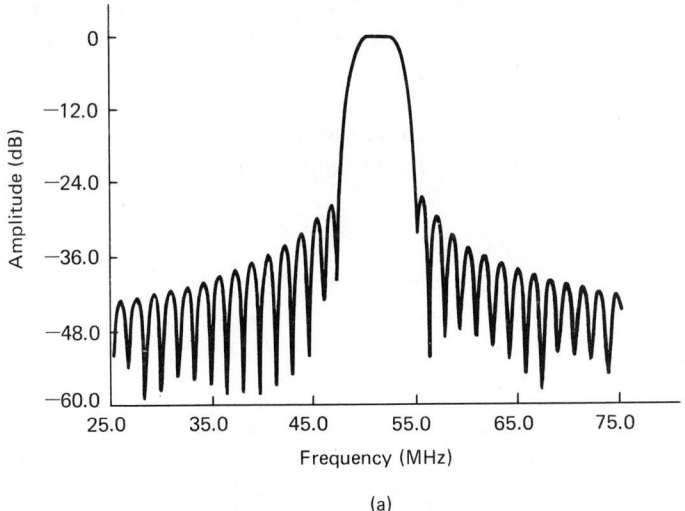

Figure 1.16 (a) Frequency response.

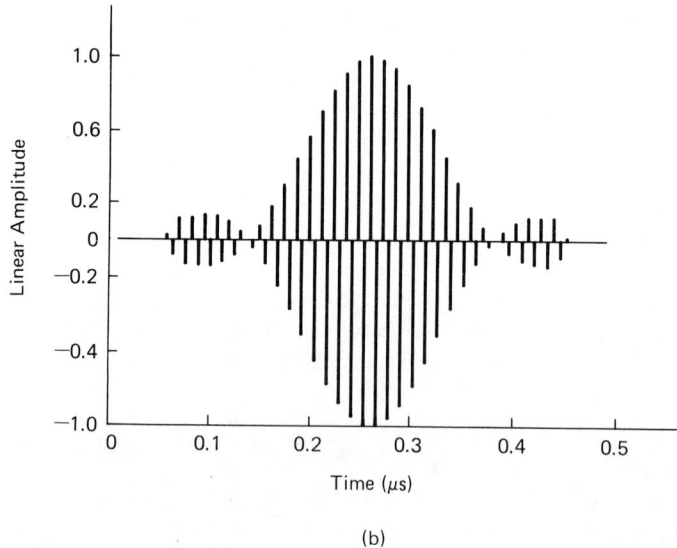

Figure 1.16 (b) corresponding impulse response.

Example 1.16
We wish to build a bandpass filter that is *not* symmetric about the center frequency. Can we use two taps per cycle?

Solution
With two taps per cycle, $F = 2f_c$ and the frequency response is periodic with period F (Section 1.2.3). This means that a frequency response that is not symmetric about f_c will look as shown in Fig. 1.17. But this is impossible as long as the tap weights are real, since $H(-f)$ must equal $H^*(f)$. Hence to implement a nonsymmetric passband, we should use four taps per cycle or use nonuniform sampling.

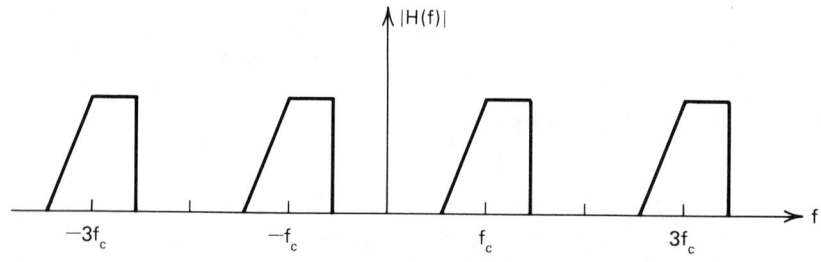

Figure 1.17 Frequency response in Example 1.16. This is physically unrealizable.

2

TRANSMISSION LINES AND PLANE ACOUSTIC WAVES

Our main objective in this chapter is to discuss the generation of plane acoustic waves in piezoelectric solids, emphasizing the analogy with the excitation of waves in a transmission line. The familiar transmission line is a perfect prototype for all one-dimensional wave propagation phenomena. As long as the power flow can be described in terms of two quantities (like voltage and current) that vary in only one spatial dimension, the problem can be reduced to that of a transmission line. In Section 2.1 we review basic transmission line theory and discuss the lumped equivalent circuit for a transmission line excited by a current source. Various representations, such as the $[Z]$, $[T]$, and $[S]$ matrices, used to describe multiport networks are also discussed in this context. The propagation of plane acoustic waves (shear and compressional) is discussed in Section 2.2, where we point out the analogy with transmission lines. After a brief introduction to piezoelectricity in Section 2.3, we go on in Section 2.4 to the generation of plane acoustic waves in piezoelectric solids. It is shown that the equations describing the generation of acoustic waves by applied electric fields are identical to those describing the excitation of waves in a transmission line by a current source. This analogy is used to derive the Mason model, which is a lumped equivalent circuit describing the generation of acoustic waves in piezoelectric solids.

For plane acoustic waves, the fields involve only a few components; consequently, it is possible to reduce the basic equations into a transmission-line form. In this book we are concerned with surface acoustic waves having many field components, and an exact reduction to the transmission line form is not possible. However, the generation of waves by an applied electric field can still be modeled *approximately* as a transmission line excited by a current source. A rigorous derivation of this result is beyond the scope of the book; instead, in Section 2.4.2 the main concept is presented using heuristic arguments and is illustrated with an example. As shown in this section, as long as the sources are purely electrical, it is particularly convenient to use the electrostatic potential, ϕ, accompanying an acoustic wave in a piezoelectric solid to denote its amplitude and phase. This is the approach we use in Chapter 3 to define the amplitude of surface waves and is used throughout the book; very little explicit use is made of concepts such as stress and strain, which are commonly used to describe acoustic waves.

An important relation obtained in this chapter is Eq. 2.58e:

$$K^2 y_0 = \omega \epsilon k$$

where $y_0 = 2P/A|\phi|^2$

P = power carried by a wave of amplitude ϕ and cross-sectional area A

K^2 = piezoelectric coupling constant (Eq. 2.43b)

$\omega = 2\pi \times$ frequency

$k = \omega/v_0 = 2\pi/$wavelength

v_0 = wave velocity

ϵ = dielectric constant

This relation will be slightly modified in Chapter 3 for surface waves.

2.1. Transmission Lines

2.1.1. Distributed LC-Circuit Representation

Waves on a transmission line are described by two field quantities, a voltage V and a current I. A transmission line is commonly modeled as a distributed LC circuit (Fig. 2.1), L being the inductance per unit length and C, the capacitance per unit length. From elementary circuit theory, the change in voltage, dV, and the change in current, dI, over a length dz are written as

$$\frac{\partial V}{\partial z} = -L \frac{\partial I}{\partial t} \qquad (2.1a)$$

2 TRANSMISSION LINES AND PLANE ACOUSTIC WAVES

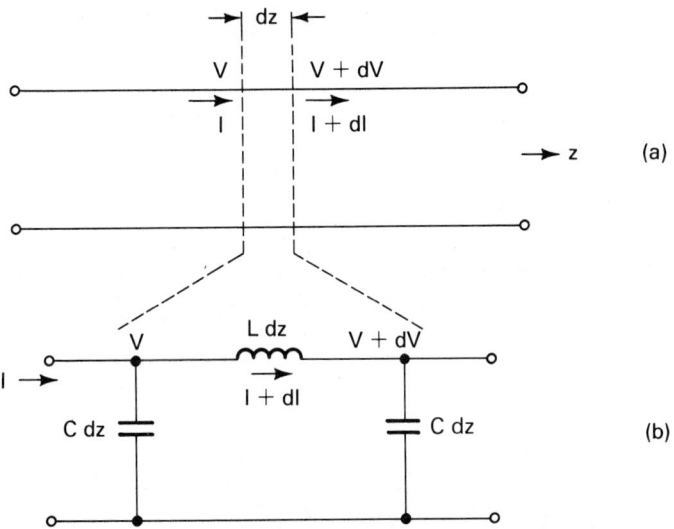

Figure 2.1 (a) Transmission line; (b) distributed LC model of a transmission line segment of length dz.

$$\frac{\partial I}{\partial z} = -C \frac{\partial V}{\partial t} \tag{2.1b}$$

Equations 2.1 are the basic transmission line equations that can be solved readily to obtain the wave velocity, v_0, and the characteristic impedance, Z_0, in terms of the distributed LC parameters. For example, eliminating I from Eq. 2.1, we have

$$\frac{\partial^2 V}{\partial z^2} = LC \frac{\partial^2 V}{\partial t^2} \tag{2.2}$$

If we assume a traveling voltage (and current) wave of the form exp $[j(\omega t - kz)]$, then $\frac{\partial}{\partial t}$ can be replaced by the multiplier $-jk$ and $\frac{\partial}{\partial t}$ by $j\omega$, so that

$$k^2 = \omega^2 LC \tag{2.3}$$

where k is the wave number related to the wavelength λ by $k = 2\pi/\lambda$ and ω is the radian frequency related to the usual frequency by $\omega = 2\pi f$. The phase velocity, v_0, is obtained from Eq. 2.3:

2 TRANSMISSION LINES AND PLANE ACOUSTIC WAVES

$$v_0 = f\lambda = \frac{\omega}{k} = \frac{1}{\sqrt{LC}} \tag{2.4}$$

From Eq. 2.1a and Eq. 2.3 we obtain the characteristic impedance Z_0:

$$Z_0 = \frac{V}{I} = \sqrt{\frac{L}{C}} \tag{2.5}$$

Conversely, from Eqs. 2.4 and 2.5 we may write L and C in terms of Z_0 and v_0:

$$L = \frac{Z_0}{v_0} \tag{2.6a}$$

$$C = \frac{1}{Z_0 v_0} \tag{2.6b}$$

It should be pointed out here that in Eq. 2.5 we are using V and I to denote the phasor amplitudes of these field quantities; to get the instantaneous values we multiply by $\exp[j(\omega t - kz)]$ and take the real part. For a wave traveling in the negative z direction, however, we should multiply by $\exp[j(\omega t + kz)]$. In some cases the z dependence is mentioned explicitly while the factor of $e^{j\omega t}$ is implied.

Example 2.1
A 50-Ω transmission line has a wave of amplitude 1 V traveling in the positive z direction and a wave of amplitude $0.5 e^{j\pi/2}$ volts traveling in the negative z direction. Calculate the instantaneous values of the voltage $v(t)$ and the current $i(t)$ along the line.

Solution

$$v(t) = \text{Re}\left[1 e^{j(\omega t - kz)} + 0.5 e^{j\pi/2} e^{j(\omega t + kz)}\right]$$

$$= \left[\cos(\omega t - kz) - 0.5 \sin(\omega t + kz)\right] \text{ volts}$$

$$i(t) = \frac{1}{Z_0} \text{Re}\left[1 e^{j(\omega t - kz)} - 0.5 e^{j\pi/2} e^{j(\omega t + kz)}\right]$$

$$= 20 \left[\cos(\omega t - kz) + 0.5 \sin(\omega t + kz)\right] \text{ mA}$$

2 TRANSMISSION LINES AND PLANE ACOUSTIC WAVES

Note the negative sign in the current for the reverse wave; for a wave traveling in the negative z direction, $V = -IZ_0$.

The time-averaged power flow P for a wave in either direction is given by

$$P = \frac{1}{2} VI^* = \frac{|V|^2}{2Z_0} \tag{2.7a}$$

where $*$ denotes complex conjugation. Thus, in this example, 20 mW of power is carried in the positive z direction and 5 mW in the negative z direction. The time-averaged stored energy U is given by

$$U = \frac{1}{2} C |V|^2 = \frac{1}{2} L |I|^2 \tag{2.7b}$$

The power flow and the stored energy for waves in either direction have a simple relationship.

$$P = Uv_0 \tag{2.8}$$

2.1.2. Lumped Circuit Equivalent for a Transmission Line Segment

We have seen in Section 2.1.1 that a transmission line can be modeled as a distributed LC circuit. Alternatively, we can also represent a transmission-line segment by the lumped circuit model shown in Fig. 2.2b. This equivalence is, of course, not obvious, and our objective in this section is to demonstrate it (Examples 2.2 and 2.3); however, let us first introduce the various representations such as $[Z]$, $[T]$, and $[S]$ matrices that are often used to describe two-port networks.

Consider a segment of a transmission line of length ℓ (Fig. 2.2). We wish to treat it as a two-port network and describe it in terms of its terminal parameters. These parameters can be chosen in many different ways. One way is to use the amplitudes V^+ and V^- of the waves traveling in the positive and negative z directions, respectively. This gives us four quantities, V_1^+, V_1^-, V_2^+, and V_2^-; any two of these can be treated as dependent variables and expressed in terms of the other two through a 2×2 matrix. For example, choosing the outgoing wave amplitudes V_1^- and V_2^+ as the dependent variables, we get the scatter matrix $[S]$:

2 TRANSMISSION LINES AND PLANE ACOUSTIC WAVES

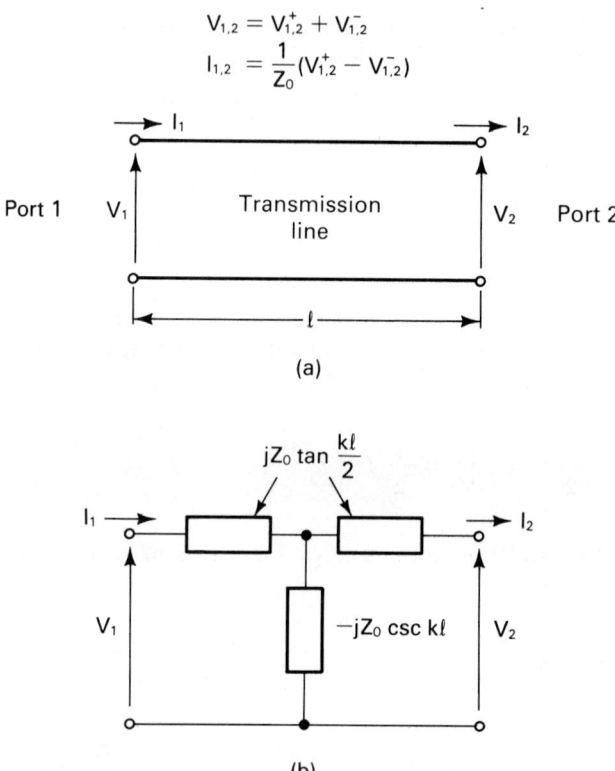

Figure 2.2 (a) Transmission-line segment of length ℓ; (b) lumped circuit equivalent.

$$\begin{Bmatrix} V_1^- \\ V_2^+ \end{Bmatrix} = \begin{bmatrix} S_{11} & S_{12} \\ S_{21} & S_{22} \end{bmatrix} \begin{Bmatrix} V_1^+ \\ V_2^- \end{Bmatrix} \tag{2.9a}$$

It is easy to see that

$$[S] = \begin{bmatrix} 0 & e^{-jk\ell} \\ e^{-jk\ell} & 0 \end{bmatrix} \tag{2.9b}$$

A linear two-port is completely described by a 2×2 matrix. However, the scatter matrix is just one of many possibilities. Different choices of dependent and independent variables lead to different matrices. For example, it is often more convenient to use the voltage V and the current I rather than the wave amplitudes V^+ and V^-.

2 TRANSMISSION LINES AND PLANE ACOUSTIC WAVES

$$V = V^+ + V^- \tag{2.10a}$$

$$I = \frac{1}{Z_0}(V^+ - V^-) \tag{2.10b}$$

Using the voltages as the dependent variables, we get the impedance matrix $[Z]$:

$$\begin{Bmatrix} V_1 \\ V_2 \end{Bmatrix} = \begin{bmatrix} Z_{11} & Z_{12} \\ Z_{21} & Z_{22} \end{bmatrix} \begin{Bmatrix} I_1 \\ I_2 \end{Bmatrix} \tag{2.11}$$

Again we could have chosen the currents as the dependent variables to get the admittance matrix $[Y]$. Another commonly used description is the transmission matrix $[T]$, which is obtained by treating the port 2 parameters (V_2, I_2 or V_2^+, V_2^-) as the dependent variables and the port 1 parameters as the independent variables. This representation is particularly useful in numerical calculations because the $[T]$ matrix for a number of sections in cascade is obtained readily by multiplying out the individual $[T]$ matrices. Given any one of these various matrices, we can calculate any other by straightforward algebraic manipulation.

We are now ready to demonstrate the equivalence between the lumped circuit of Fig. 2.2b and the transmission-line segment of Fig. 2.2a; this is done by showing that their $[Z]$ matrices are equal (see Examples 2.2 and 2.3).

Example 2.2
Calculate the $[T]$ and $[Z]$ matrices for the transmission-line section in Fig. 2.2.

Solution
We have, from Eqs. 2.9,

$$V_1^- = e^{-jk\ell} V_2^-$$

$$V_2^+ = e^{-jk\ell} V_1^+$$

Hence

$$V_2 = V_2^+ + V_2^-$$

$$= e^{-jk\ell} V_1^+ + e^{jk\ell} V_1^-$$

$$= V_1 \cos k\ell - j Z_0 I_1 \sin k\ell$$

Similarly,

$$Z_0 I_2 = V_2^+ - V_2^-$$
$$= Z_0 I_1 \cos k\ell - j V_1 \sin k\ell$$

This gives us the transmission matrix:

$$\begin{Bmatrix} V_2 \\ I_2 \end{Bmatrix} = \begin{bmatrix} \cos k\ell & -jZ_0 \sin k\ell \\ \dfrac{-j \sin k\ell}{Z_0} & \cos k\ell \end{bmatrix} \begin{Bmatrix} V_1 \\ I_1 \end{Bmatrix} \qquad (2.12)$$

We can now get the $[Z]$ matrix with some simple algebraic manipulation:

$$\begin{Bmatrix} V_1 \\ V_2 \end{Bmatrix} = Z_0 \begin{bmatrix} -j \cot k\ell & j \csc k\ell \\ -j \csc k\ell & j \cot k\ell \end{bmatrix} \begin{Bmatrix} I_1 \\ I_2 \end{Bmatrix} \qquad (2.13)$$

Here we have chosen the reference direction for both I_1 and I_2, in the positive Z-direction. Sometimes the reference direction for I_2 is reversed; the signs of the matrix elements are then appropriately modified.

Example 2.3
Calculate the $[Z]$ matrix for the circuit shown in Fig. 2.2b.

Solution

$$Z_{11} = \left. \frac{V_1}{I_1} \right|_{I_2 = 0} = -jZ_0 \left[\csc k\ell - \tan \frac{k\ell}{2} \right]$$

$$Z_{21} = \left. \frac{V_2}{I_1} \right|_{I_2 = 0} = -jZ_0 \csc k\ell$$

Similarly, we can get Z_{12} and Z_{22}. Note that the $[Z]$ matrix for this lumped circuit is the same as that for the transmission-line segment (Eq. 2.13), thus demonstrating their equivalence.

2.1.3. Generation of Waves by a Current Source

Let us now consider a transmission line excited by a distributed current source I_s per unit length (Fig. 2.3a). Equations 2.1 are now slightly modified to include I_s:

2 TRANSMISSION LINES AND PLANE ACOUSTIC WAVES

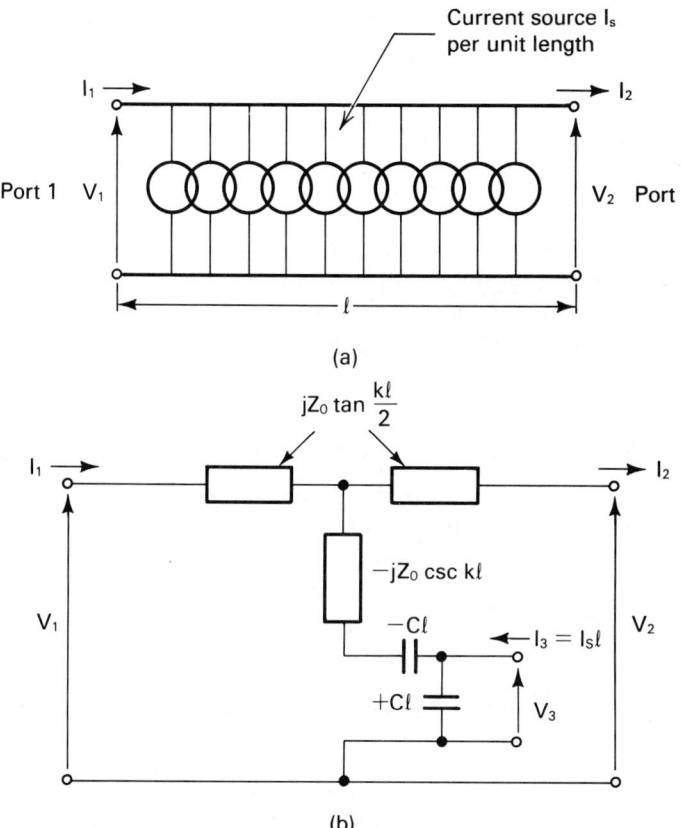

Figure 2.3 (a) Transmission-line segment of length ℓ driven by a constant-current source I_s per unit length; (b) lumped circuit equivalent.

$$\frac{\partial V}{\partial z} = -L \frac{\partial I}{\partial t} \tag{2.14a}$$

$$\frac{\partial I}{\partial z} = -C \frac{\partial V}{\partial t} + I_s \tag{2.14b}$$

In this section we wish to show that the three-port lumped circuit in Fig. 2.3b can be used to represent the transmission line excited by a constant I_s. In general, I_s is a function of z and it is fairly straightforward to calculate the waves generated in either direction by the source (see Section 4.2). For the moment, however, let us assume a constant I_s independent of z to keep the algebra simple.

To describe the generation of waves by the source current, we can extend the scatter matrix in Eqs. 2.9:

$$\begin{Bmatrix} V_1^- \\ V_2^+ \end{Bmatrix} = \begin{bmatrix} 0 & e^{-jk\ell} & Z_m \\ e^{-jk\ell} & 0 & Z_m \end{bmatrix} \begin{Bmatrix} V_1^+ \\ V_2^- \\ I_s \end{Bmatrix} \quad (2.15\text{a})$$

The last column of this matrix describes the generation of outgoing wave amplitudes V_1^- and V_2^+ by the source I_s. It is easy to calculate the parameter Z_m describing the coupling to the source. A mathematical derivation is given in Example 2.4 below. However, we can get the same result with a simple argument as follows. The current source in a short element of length dz generates waves of amplitude $\frac{1}{2} Z_0 I_s \, dz$ in either direction. The total amplitude is obtained by summing all the waves generated in the region $z = 0$ to $z = \ell$.

$$Z_m = \frac{Z_0}{2} \int_0^\ell e^{jk(z-\ell)} \, dz$$

$$= \frac{jZ_0}{2k} \left(e^{-jk\ell} - 1 \right) \quad (2.15\text{b})$$

The factor of $e^{jk(z-\ell)}$ comes in because waves generated at different points have to be shifted to a common reference plane $(z = \ell)$ before summing. This is described in more detail in Section 4.2.

Example 2.4
Derive an expression for Z_m starting from the transmission-line equations (Eqs. 2.14). Combining Eqs. 2.14a and 2.14b and assuming an $\exp(j\omega t)$ time dependence,

$$\frac{\partial^2 V}{\partial z^2} + k^2 V = -j\omega L I_s, \qquad k = \frac{\omega}{v_0} \quad (2.16\text{a})$$

Solution
If the right-hand side were zero, we would have (for waves traveling in the $+z$ direction)

$$V(z) = A e^{-jkz} \quad (2.16\text{b})$$

2 TRANSMISSION LINES AND PLANE ACOUSTIC WAVES

where A is a constant. In the present case, because of the source term, we expect A to grow with z. Substituting Eq. 2.16b in Eq. 2.16a, we get an equation describing the growth of A.

$$\frac{\partial^2 A}{\partial z^2} - 2jk \frac{\partial A}{\partial z} = -j\omega L I_s e^{jkz} \tag{2.17a}$$

If A grows slowly so that the change in A is very small over a wavelength, we can neglect the first term on the left relative to the second term. With this approximation we have

$$\frac{\partial A}{\partial z} = \frac{Z_0 I_s}{2} e^{jkz} \tag{2.17b}$$

$$A(z) = A(0) + \frac{Z_0 I_s}{2} \int_0^z dz' \, e^{jkz'}$$

Using Eq. 2.16b yields

$$V(z) = V(0) + \frac{Z_0 I_s}{2} \int_0^z dz' \, e^{jk(z'-z)} \tag{2.18}$$

If $V(0) = 0$,

$$V(\ell) = \frac{Z_0 I_s}{2} \int_0^\ell dz \, e^{jk(z-\ell)}$$

This leads to the expression for Z_m in Eq. 2.15b.

As before, we can change our dependent and independent variables as convenient. For example, carrying out the transformation described by Eqs. 2.10, we get

$$\begin{Bmatrix} V_2 \\ I_2 \end{Bmatrix} = \begin{bmatrix} \cos k\ell & -jZ_0 \sin \ell & Z_m(1-e^{jk\ell}) \\ \dfrac{-j \sin k\ell}{Z_0} & \cos k\ell & \dfrac{Z_m}{Z_0}(1+e^{jk\ell}) \end{bmatrix} \begin{Bmatrix} V_1 \\ I_1 \\ I_s \end{Bmatrix} \tag{2.19}$$

Also,

$$\begin{Bmatrix} V_1 \\ V_2 \end{Bmatrix} = Z_0 \begin{bmatrix} -j \cot k\ell & j \csc k\ell & \dfrac{-j}{k\ell} \\ -j \csc k\ell & j \cot k\ell & \dfrac{-j}{k\ell} \end{bmatrix} \begin{Bmatrix} I_1 \\ I_2 \\ I_s \end{Bmatrix} \tag{2.20}$$

In this discussion we have used an idealized current generator I_s for the source. In practice, the source should be described by a voltage V_s and a current I_s. This would introduce another independent variable, V_s, leading to a 3×3 matrix rather than a 2×3 matrix. This is understandable, considering that we now have a three-port instead of a two-port. However, to generate the third row of this matrix, we have to consider the detailed nature of the source and its coupling. In Section 2.3 we will see how a plane acoustic wave in a piezoelectric solid coupled to an electrical source is described by a 3×3 matrix.

Example 2.5
Calculate the $[Z]$ matrix for the three-port lumped network in Fig. 2.3b and show that it is identical to the $[Z]$ matrix for the transmission line (Eq. 2.20).

Solution

$$[Z] = Z_0 \begin{bmatrix} -j \cot k\ell & j \csc k\ell & -\dfrac{j}{k\ell} \\ -j \csc k\ell & j \cot k\ell & -\dfrac{j}{k\ell} \\ -\dfrac{j}{k\ell} & +\dfrac{j}{k\ell} & -\dfrac{j}{k\ell} \end{bmatrix} \quad (2.21)$$

Here we have used the relation $CZ_0 = \dfrac{1}{v_0} = \dfrac{k}{\omega}$. It is easy to see that the first two rows in Eq. 2.21 are identical to Eq. 2.20, noting that $I_3 = I_s \ell$ is the total current from the current generator.

2.2. Uniform Plane Acoustic Waves

2.2.1. Compressional Waves

We will now describe a simple kind of acoustic wave, namely, a plane uniform compressional wave, and show its similarity to the electrical wave in a transmission line. We first need to define some acoustic field quantities which take the place of voltages and currents. Let us consider a simple experiment in which a rod, initially of length L, is stretched by an amount when a force F is applied uniformly over its cross section, A (Fig. 2.4). This is a case of static stress and does not involve any wave motion, but it provides a simple illustration for some of the field quantities that come up in the description of acoustic waves.

2 TRANSMISSION LINES AND PLANE ACOUSTIC WAVES

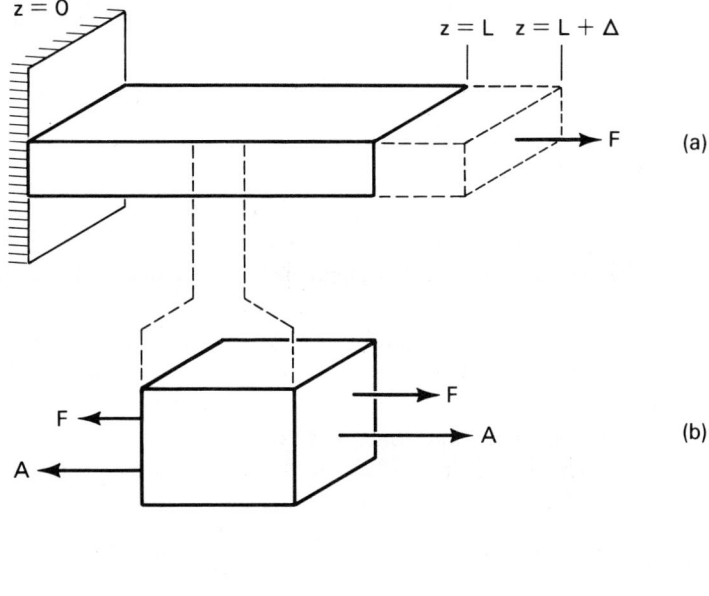

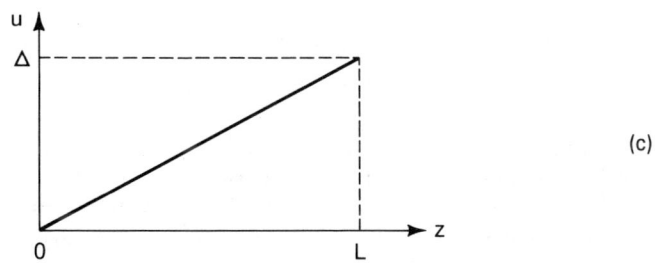

Figure 2.4 (a) Uniform dilation of a rod by an applied force; (b) section of the rod showing the forces on the different faces; (c) particle displacements at various points along the rod.

The stress T is defined as the force exerted per unit area. If we consider a section of the rod (Fig. 2.4b), we note that the force is equal and opposite on its two faces; in fact, that is what keeps it in equilibrium. This might give the impression that the stress is positive on the right-hand face and negative on the left-hand face. However, we have to note that the area is also defined as a vector; its direction is that of the outward going normal from the volume under consideration. Thus both force and area reverse sign on the left-hand face, making the stress positive on either face. The stress is positive for stretching (or dilation) and negative for compression.

The stress is usually given two subscripts, the first denoting the direction of the force and the second denoting the direction of the area:

$$T_{ij} = \frac{F_i}{A_j} \qquad i, j = x, y, z \tag{2.22}$$

In this example, both F and A are in the z direction, so that the stress is denoted as T_{zz}. Stresses for which the force and area are in different directions (such as T_{yz}) are called shearing stresses and will be discussed later.

Now we come to the concept of strain, S. Strain is the fractional change in length, that is, $S = \Delta/L$. A more useful definition of strain is obtained by considering a plot of the particle displacement, u (from the unstrained condition) as a function of distance along the rod (Figure 2.4c). The left end of the rod is rigidly fixed so that its displacement is zero, while the right end is displaced by an amount Δ. The displacement varies linearly in between. The strain is defined as the rate of change of displacement in a particular direction:

$$S = \frac{\partial u}{\partial z}$$

We note that this definition is equivalent to Δ/L for the case of uniform strain considered here, but for nonuniform strains this is a more useful definition.

The strain, in general, is also defined with two subscripts since the particle displacement can have one of three directions and its rate of change can be considered in one of three directions. In this case since the particle is displaced in the z direction and we are considering its change in the z direction, the strain is written as S_{zz}. The two subscripts are different for shearing strains, which will be considered later.

Now let us consider a compressional wave traveling in the z direction in a block of material (Fig. 2.5). The particles in this type of wave are displaced in the direction of wave propagation, so that compressional waves are also known as longitudinal waves. We are now considering a wave rather than a static stress; the particle displacement u_z varies with time and distance as $\exp[j(\omega t - kz)]$. A uniform plane wave is assumed so that there is no variation in fields in the x or y direction. We can define a particle velocity v (not to be confused with the wave velocity v_0) as the time derivative of u.

$$v_z = \frac{\partial u_z}{\partial t} \tag{2.23a}$$

The strain is the space derivative of u (for infinitesimal strains).

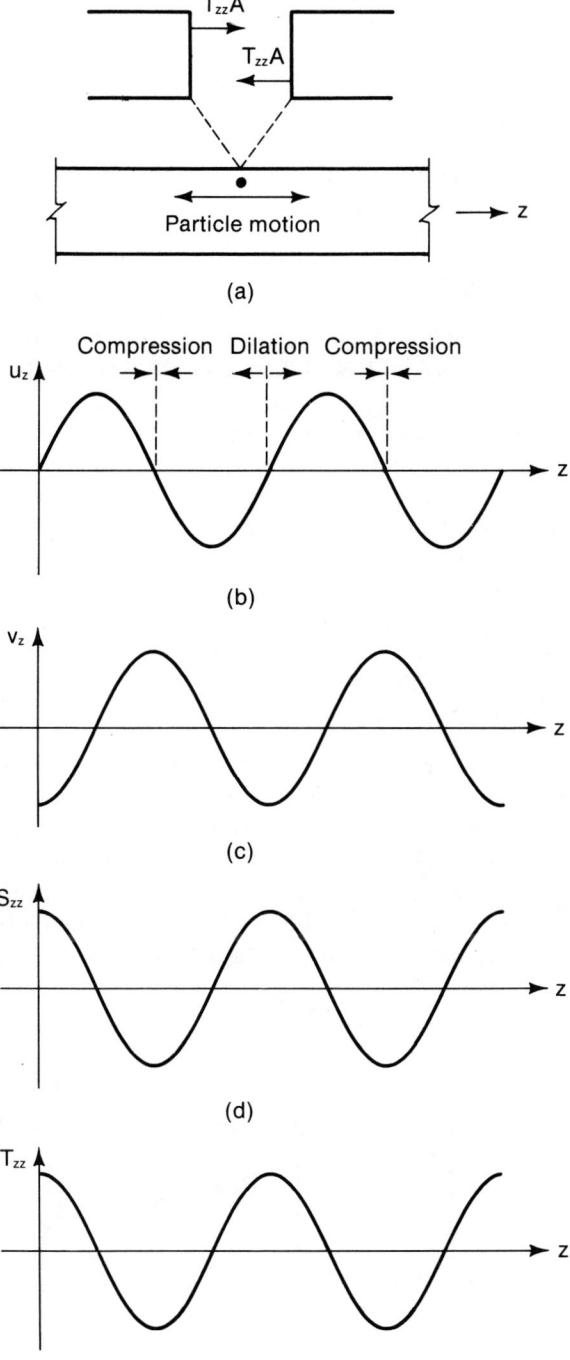

Figure 2.5 (a) Particle motion and stress generated by compressional (longitudinal) wave; (b) u_z versus z; (c) v_z versus z; (d) S_{zz} versus z; (e) T_{zz} versus z.

$$S_{zz} = \frac{\partial u_z}{\partial z} \qquad (2.23\text{b})$$

The laws of elasticity state that for small strains, the stress is proportional to the strain.

$$T_{zz} = c_{zzzz} S_{zz} \qquad (2.24)$$

where c is known as the stiffness coefficient or the elasticity tensor. (This linearity, of course, does not hold for large enough strains; in fact, the operation of certain SAW devices such as convolvers is based on the nonlinear relationship between stress and strain.) Since there are various kinds of possible stresses and strains, we need the four subscripts on c to indicate that it relates a compressional strain in the z direction (last two subscripts) with the compressional stress in the z direction (first two subscripts).

A little manipulation with Eqs. 2.23 and 2.24 yields

$$\frac{\partial v_z}{\partial z} = \frac{1}{c_{zzzz}} \frac{\partial T_{zz}}{\partial t} \qquad (2.25\text{a})$$

We note that this is very similar to the transmission-line equations (Eqs. 2.1), with the particle velocity v_z and the stress T_{zz} replacing the current and voltage. To complete the analogy, however, we need another equation that relates $\partial T_{zz}/\partial z$ with $\partial v_z/\partial t$. This equation is obtained by applying Newton's second law to a differential volume $dx\ dy\ dz$ of the block (Fig. 2.6). The net force on the volume depends on the change of T_{zz} from the left face to the right $[dz\ (\partial T_{zz}/\partial z)]$. This is equated to the mass of the volume times its acceleration:

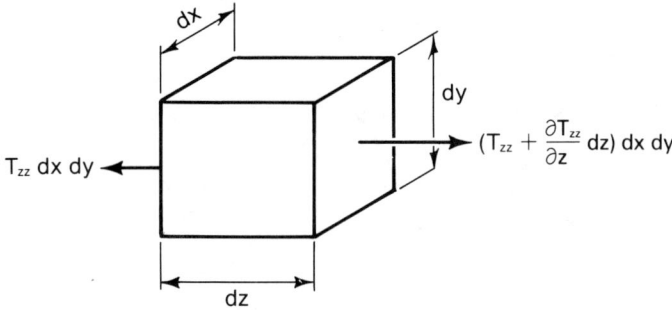

Figure 2.6 Forces acting on a differential volume.

2 TRANSMISSION LINES AND PLANE ACOUSTIC WAVES

$$\frac{\partial T_{zz}}{\partial z}\, dz\, dx\, dy = (\rho\, dz\, dx\, dy)\, \frac{\partial v_z}{\partial t}$$

ρ is the mass density of the material. Canceling out the volume element gives us

$$\frac{\partial T_{zz}}{\partial z} = \rho\, \frac{\partial v_z}{\partial t} \qquad (2.25\text{b})$$

Together with Eq. 2.25a, this completes the analogy with the transmission-line equations (Eqs. 2.1). In fact, if we equate T_{zz} to voltage and v_z to current (the reverse is just as logical), we note that the inductance L is replaced by the density ρ and the capacitance C is replaced by c^{-1}, the reciprocal of the stiffness (dropping the subscripts for convenience). Clearly, then, the wave velocity is given by

$$v_o = \sqrt{\frac{c}{\rho}} \qquad (2.26)$$

It will be noted that the acoustic equations 2.25 differ from the transmission line equations 2.1 by a negative sign. The negative sign may be removed by relating $-T_{zz}$ (rather than $+T_{zz}$) to the voltage V. This makes no difference to the velocity, but it does introduce a negative sign in the power flow.

$$Z_0 = \frac{-T_{zz}}{v_z} = \sqrt{c\rho} \qquad (2.27)$$

$$P = -\frac{1}{2}\, v_z^* T_{zz} \qquad (2.28)$$

Equation 2.28 yields the power flow per unit area and has to be multiplied by the cross-sectional area to obtain the total power carried by the wave. Again, from the transmission-line analogy we have a definition for the stored energy per unit volume U (see Eq. 2.7b).

$$U = \frac{|T|^2}{2c} = \frac{\rho |v|^2}{2} \qquad (2.29)$$

(We have dropped all subscripts for convenience.) It can be easily checked that U and P are still related by Eq. 2.8, namely, $P = Uv_0$.

Example 2.6

Consider a solid with $C_{zzzz} = 10^{11}$ N/m² and $\rho = 1000$ kg/m³. A compressional wave at 1 MHz carrying 500 W/m² is propagating in the positive z direction. Calculate T_{zz}, v_z, u_z, S_{zz}, and U.

Solution

To take care of the units in this problem, it is useful to note that

$$N = kg \cdot m/s^2$$

$$W = N \cdot m/s$$

$$P = 500 \text{ W/m}^2$$

$$v_0 = 10^4 \text{ m/s} \tag{2.26}$$

$$Z_0 = 10^7 \text{ N} \cdot \text{s/m}^3 \tag{2.27}$$

$$f = 1 \text{ MHz}$$

$$\lambda = \frac{v_0}{f} = 10^{-2} \text{ m}$$

$$T_{zz} = -\sqrt{2PZ_0} = -10^5 \text{ N/m}^2 \tag{2.27}, (2.28)$$

$$v_z = 10^{-2} \text{ m/s} \tag{2.27}$$

$$u_z = 16j\text{Å}$$

$$S_{zz} = 10^{-6} \tag{2.23b}, (2.24)$$

$$U = 0.05 \text{ J/m}^3 \tag{2.29}$$

Note that $U = P/v_0 = S_{zz}^* T_{zz}/2$.

2.2.2. Shear Waves

Consider the same situation as in Fig. 2.4a but with the force applied upward instead of sideways (Fig. 2.7a). The rod is now strained but with no change in volume. There is no compression or dilation since adjacent particles do not push each other: rather, they tend to slide against each

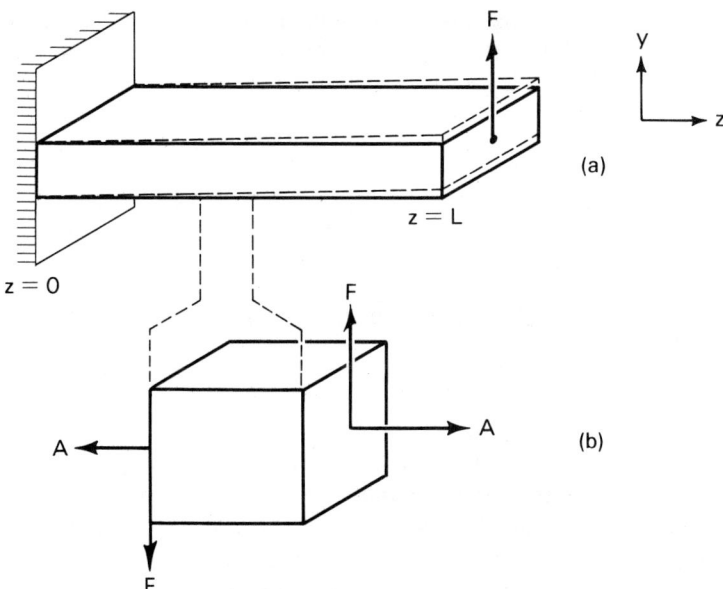

Figure 2.7 (a) Shear straining of a rod by an applied force; (b) section of the rod showing the forces on the different faces.

other, producing what are called shearing stresses and strains. The shearing stress is readily obtained using Eq. 2.22 (Fig. 2.7).

$$T_{yz} = \frac{F}{A}$$

In this case, too, the particle displacement varies linearly along the rod as in the compressional case (Fig. 2.6); however, the displacement is transverse to the rod and not along it. By analogy, it seems that we should define shearing strain as $S_{yz} = \partial u_y/\partial z$. However, this definition is not appropriate. If we merely rotate a rod without straining it (as would be the case if the rod in Fig. 2.6 were not fixed rigidly at $z=0$), $\partial u_y/\partial z$ is nonzero; so it is not really a good measure of the strain.

At this point let us digress a little to discuss something known as the *abbreviated subscript notation*. We have seen that the stress is specified by two subscripts, the first denoting the direction of the force and the second denoting the direction of the area. Since there are three directions we can visualize nine components of stress by combining each possible force direction with each possible area direction. However, it can be shown that interchanging the subscripts leaves the stress unchanged; for example, $T_{yz} = T_{zy}$ (this is not meant to be obvious, but we shall not prove it here). This leaves only six independent components of stress. The abbreviated subscript notation takes advantage of this fact to replace the double subscript on T with a single subscript, using the following prescription:

xx	1
yy	2
zz	3
yz, zy	4
zx, xz	5
xy, yx	6

When discussing a compressional wave propagating in the z direction we encountered a stress component T_{zz}; in the abbreviated notation this is written as T_3. In our present example we have the stress component T_{yz}, which is abbreviated as T_4. A similar abbreviation of subscripts is also possible for the strain, S_4. The shearing strains are defined as follows. This definition eliminates pure rotations (this is not meant to be obvious either).

$$S_4 = \frac{\partial u_z}{\partial y} + \frac{\partial u_y}{\partial z}$$

$$S_5 = \frac{\partial u_x}{\partial z} + \frac{\partial u_z}{\partial x} \qquad (2.30)$$

$$S_6 = \frac{\partial u_y}{\partial x} + \frac{\partial u_x}{\partial y}$$

The compression strains S_1, S_2, and S_3 are the same as S_{xx}, S_{yy}, and S_{zz} defined in connection with compressional waves (Eq. 2.23b). Since both T and S now have a single subscript (that runs from 1 through 6), the stiffness coefficient has two subscripts: the first denoting the stress and the second, the strain. Thus c_{zzzz} (see Eq. 2.24) is replaced by c_{33}; the velocity of a compressional wave in the z direction is $(c_{33}/\rho)^{1/2}$.

So far we have talked of static shearing stresses and strains. Next, let us consider an acoustic wave propagating in the z direction but with its particle displacements in the y direction (or the x direction) (Fig. 2.8). Such a wave is called a shear (or transverse) wave. A y-polarized, z-propagating shear wave produces a stress T_{yz}, which in our new notation is T_4. Looking at Eq. 2.30, we note that the only nonzero strain is S_4. This is because $u_z = u_x = 0$ and $\partial/\partial x = \partial/\partial y = 0$ (since we are considering a z-propagating uniform plane wave).

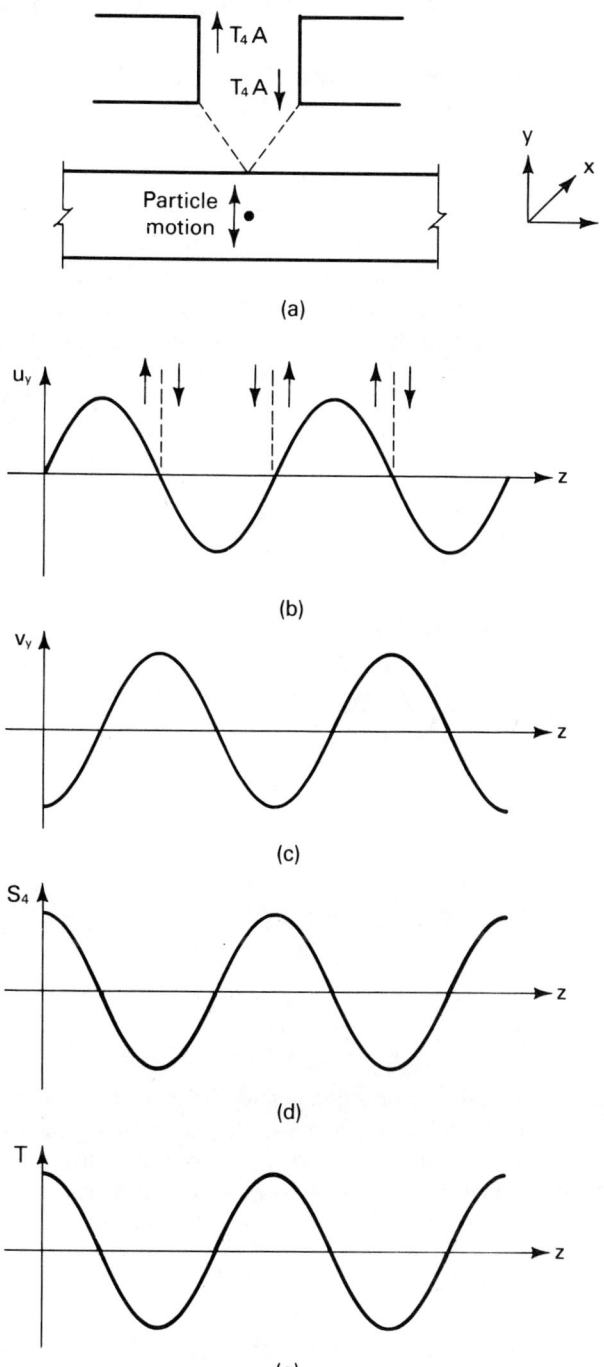

Figure 2.8 (a) Particle motion and stress generated by a y-polarized, z-propagating shear (transverse) wave; (b) u_y versus z; (c) v_y versus z; (d) S_4 versus z; (e) T_4 versus z.

2 TRANSMISSION LINES AND PLANE ACOUSTIC WAVES

The stress and strain are related through a stiffness coefficient, c_{44}, as in Eq. 2.24.

$$T_4 = c_{44} S_4 \tag{2.31}$$

Comparing with the compressional wave discussed earlier, we note that we have essentially the same equations with u_z, v_z replaced by u_y, v_y; $T_{zz}(T_3)$ replaced by $T_{yz}(T_4)$; and c_{33} replaced by c_{44}. (Compare Figs. 2.5 and 2.8.) In fact, transmission-line equations similar to Eq. 2.25 are readily derived:

$$\frac{\partial v_y}{\partial z} = \frac{1}{c_{44}} \frac{\partial T_{yz}}{\partial z} \tag{2.32a}$$

$$\frac{\partial T_{yz}}{\partial z} = \rho \frac{\partial v_y}{\partial t} \tag{2.32b}$$

As before, from the analogy with transmission lines,

$$\text{velocity}, \ v_0 = \sqrt{\frac{c_{44}}{\rho}} \tag{2.33}$$

$$\frac{\text{power}}{\text{area}}, \ P = -\frac{1}{2} v_y^* T_{yz} \tag{2.34}$$

$$\frac{\text{energy}}{\text{volume}}, \ v = \frac{|T_{yz}|^2}{2c_{44}} = \frac{\rho |v_y|^2}{2} \tag{2.35}$$

$$Z_0 = -\frac{T_{yz}}{v_y} = \sqrt{c_{44}\rho} \tag{2.36}$$

We have discussed compressional and shear acoustic waves. These are uniform plane waves and in each case only a particular component of the particle velocity v and of the stress T is nonzero; this makes possible a straightforward identification of voltage and current. This will not be possible with the more complicated field patterns associated with surface waves.

Example 2.7
Consider a solid with $C_{44} = 6.56 \times 10^{10}$ N/m^2 and $\rho = 1000$ kg/m^3. A y-polarized shear wave at 1 MHz carrying 500 W/m^2 is propagating in the positive z direction. Calculate T_4, v_y, u_y, S_4, and U.

2 TRANSMISSION LINES AND PLANE ACOUSTIC WAVES

Solution
The problem proceeds in exactly the same manner as for Example 2.6.

$$P = 500 \text{ W/m}^2$$

$$v_0 = 8100 \text{ m/s}$$

$$Z_0 = 8.1 \times 10^6 \text{ N} \cdot \text{s/m}^3$$

$$f = 1 \text{ MHz}$$

$$\lambda = 8.1 \times 10^{-3} \text{ m}$$

$$T_4 = 9 \times 10^4 \text{ N/m}^2$$

$$v_y = -1.1 \times 10^{-2} \text{ m/s}$$

$$u_y = 17.5j \text{ Å}$$

$$S_4 = 1.37 \times 10^{-6}$$

$$U = 0.062 \text{ J/m}^3$$

Note that $U = P/v_0 = \frac{1}{2} S_4^* T_4$.

Now that we have discussed the nature of acoustic fields, we are ready to introduce the concept of piezoelectricity. However, before we proceed further, let us discuss briefly the nature of the stiffness matrix c that relates stress to strain. So far we have encountered two components of c: c_{33} and c_{44}. However, since each subscript can run from 1 through 6, we note that c has $6 \times 6 = 36$ components relating the six components of stress to the six strain components. Of these, three are independent for a crystal with cubic symmetry, and only two are independent for an isotropic solid. With the x, y, and z axes oriented along the crystal axes in a *cubic* crystal, the shear and compressional effects are decoupled from each other. Since subscripts 1 through 3 denote compressional (or dilation) effects while subscripts 4 through 6 denote shearing effects, this means that the $[c]$ matrix reduces to two 3×3 matrices, one relating the compressional effects and the other relating the shearing effects:

$$\begin{Bmatrix} T_1 \\ T_2 \\ T_3 \\ T_4 \\ T_5 \\ T_6 \end{Bmatrix} \begin{matrix} \text{Com-} \\ \text{pressional} \\ = \\ \text{Shear} \end{matrix} \begin{bmatrix} c_{11} & c_{12} & c_{12} & & & \\ c_{12} & c_{11} & c_{12} & & \text{Zero} & \\ c_{12} & c_{12} & c_{11} & & & \\ & & & c_{44} & 0 & 0 \\ & \text{Zero} & & 0 & c_{44} & 0 \\ & & & 0 & 0 & c_{44} \end{bmatrix} \begin{Bmatrix} S_1 \\ S_2 \\ S_3 \\ S_4 \\ S_5 \\ S_6 \end{Bmatrix} \quad (2.37)$$

The shears in the three directions are decoupled from each other, so that the lower (3 × 3) matrix is diagonal, with each diagonal element equal to c_{44}. The compressional effects in the three directions are, however, coupled to each other through c_{12}. This coupling occurs because compressing a material in one direction causes it to bulge out in other directions.

For *isotropic* solids, only two of the constant c_{11}, c_{12}, c_{44} are independent; they are related through the relation

$$c_{11} = c_{12} + 2c_{44} \quad (2.38)$$

Isotropic solids are often specified using what are known as the Lamé constants, $\lambda \ (= c_{12})$ and $\mu \ (= c_{44})$. From Eq. 2.38, c_{11} is evidently equal to $\lambda + 2\mu$.

2.3. Piezoelectricity

SAW technology relies on the use of piezoelectric crystals in which acoustic fields give rise to electric fields, and vice versa. We will not go into the atomic details of crystal structure that cause this effect. Rather, we will discuss this in terms of the phenomenonological constants that relate the acoustic and electric fields. We have seen that acoustic fields are described in terms of stresses and strains related by

$$T = cS$$

(We have dropped the subscripts for clarity.) The corresponding quantities for electric fields are the electrical displacement D and the electric field E. They are related by

$$D = \epsilon E$$

where ϵ is the permittivity. As we know, ϵ for a dielectric is different from that of vacuum (ϵ_0) because electric fields produce polarization charges in the material. In piezoelectric solids there is an asymmetry in the arrangement of the atomic dipoles such that straining it also causes polarization charges to appear. The result is that the two fields (strain and electric) are coupled through a piezoelectric constant e:

$$T = cS - e^T E \tag{2.39a}$$

$$D = eS + \epsilon E \tag{2.39b}$$

These are called the *constitutive relations*. It should be noted that D and E are both vectors with three components each. The piezoelectric constant, $[e]$, thus has $3 \times 6 = 18$ components and the permittivity, $[\epsilon]$, has $3 \times 3 = 9$ components. The superscript T in Eq. 2.39a indicates transpose; $[e]$ is a 3×6 matrix while $[e^T]$ is a 6×3 matrix. Many of these 18 components of $[e]$ are often zero, depending on the symmetry properties of the crystal. For example, in cubic-compound semiconductors such as GaAs, the nonzero elements of $[e]$ are indicated by crosses ($\times$):

$$[e] \rightarrow \begin{bmatrix} 0 & 0 & 0 & \times & 0 & 0 \\ 0 & 0 & 0 & 0 & \times & 0 \\ 0 & 0 & 0 & 0 & 0 & \times \end{bmatrix}$$

This means that shear strains $S_{yz}(S_4)$, $S_{zx}(S_5)$, and $S_{xy}(S_6)$ give rise to electric fields E_x, E_y, and E_z, respectively. Note the symmetry among the coordinates x, y, and z; this is expected since in cubic solids x, y, and z are completely equivalent. On the other hand, the semiconductor ZnO has a preferred axis (z); in ZnO compressional strains $S_{xx}(S_1)$, $S_{yy}(S_2)$, and $S_{zz}(S_3)$ all give rise to electric fields E_z, while shear strains $S_{yz}(S_4)$ and $S_{zx}(S_5)$ give rise to electric fields E_y and E_x, respectively. ZnO is often deposited as a thin film on nonpiezoelectric substrates such as silicon in order to generate acoustic waves through the piezoelectric effect.

The common substrates used in SAW technology are lithium niobate, lithium tantalate, and quartz. The first two have $[e]$ matrices of the form

$$[e] \rightarrow \begin{bmatrix} 0 & 0 & 0 & 0 & \times & \times \\ \times & \times & 0 & \times & 0 & 0 \\ \times & \times & \times & 0 & 0 & 0 \end{bmatrix}$$

while the $[e]$ matrix for quartz has the form

$$[e] \rightarrow \begin{bmatrix} \times & \times & 0 & \times & 0 & 0 \\ 0 & 0 & 0 & 0 & \times & \times \\ 0 & 0 & 0 & 0 & 0 & 0 \end{bmatrix}$$

2.4. Generation of Acoustic Waves in Piezoelectric Solids

2.4.1. Mason Model

Let us now consider the compressional wave discussed earlier in Section 2.2.1, but assume that we have a piezoelectric solid with a nonzero e_{33} so that the compressional strain $S_{zz}(S_3)$ gives rise to an electric field E_z. Intuitively, we would expect that if we apply an electric field E_z externally (Fig. 2.9a), compressional waves will be generated. In this section we will show that the equations describing this generation process are identical to those describing a transmission line excited by a distributed current source. Consequently, the lumped equivalent circuit shown (known as the in-line Mason model) in Fig. 2.9b can be used to model the process; this is similar to Fig. 2.3b expect for a transformer that converts electrical to acoustic quantities.

For a compressional wave traveling in the z direction in a piezoelectric solid with a nonzero e_{33}, the nonzero acoustic fields are T_3 and S_3, while the nonzero electrical fields are D_3 and E_3. Hence the constitutive relations (Eqs. 2.39) can be written as

$$T_3 = c_{33}S_3 - e_{33}E_3 \tag{2.40a}$$

$$D_3 = e_{33}S_3 + \epsilon_{33}E_3 \tag{2.40b}$$

As before (Eq. 2.25b), we get from Newton's law,

$$\frac{\partial T}{\partial z} = \rho \frac{\partial v}{\partial t} \tag{2.41a}$$

(We are dropping the subscripts for convenience.) This is the first of our transmission-line equations. To get the second one, we take the partial time derivative of Eq. 2.40a:

$$\frac{\partial v}{\partial z} = \frac{1}{c}\frac{\partial T}{\partial t} + \frac{e}{c}\frac{\partial E}{\partial t} \tag{2.41b}$$

Here we have used Eqs. 2.23 to get $\partial S/\partial t = \partial v/\partial z$. Note that Eqs. 2.41 look very similar to the equations describing a transmission line excited by a current source I_s (see Eqs. 2.14). We will now replace the source term

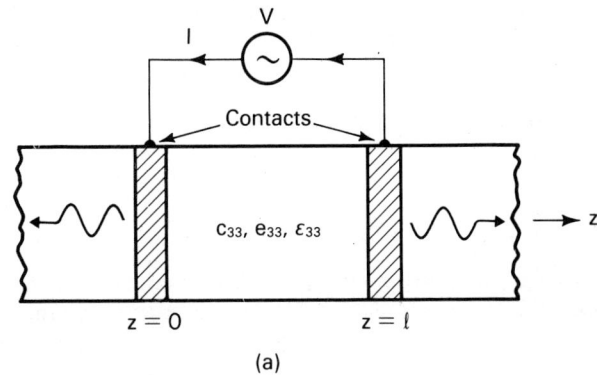

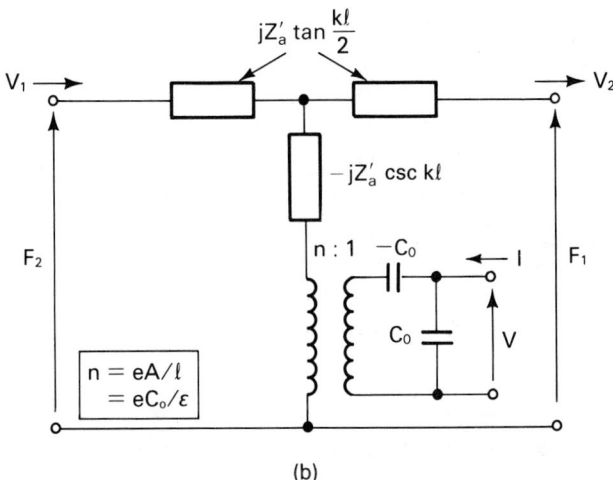

Figure 2.9 (a) Compressional wave in a piezoelectric solid coupled to an in-line electric field; (b) lumped equivalent circuit: in-line Mason model.

$(e/\epsilon)(\partial E/\partial t)$ in terms of the current I (Fig. 2.9a). Using Eq. 2.40b, we can write Eq. 2.41b as

$$\frac{\partial v}{\partial z} = \frac{1}{c'}\frac{\partial T}{\partial t} + \frac{e}{\epsilon c'}\frac{\partial D}{\partial t} \qquad (2.42)$$

where

$$c' = c + \frac{e^2}{\epsilon}$$

$$= c(1 + K^2) \qquad (2.43a)$$

$$K^2 = \frac{e^2}{\epsilon c} \qquad (2.43b)$$

We know from Poisson's equation that

$$\frac{\partial D_3}{\partial z} = \rho \qquad (D_1, D_2 = 0)$$

where ρ is the charge density. If the solid is insulating so that there are no free charges, then $\rho = 0$ everywhere in the material and D_3 must be constant. At the contacts surface charges appear equal in magnitude to D_3; the current density J in the circuit is equal to the displacement current in the material.

$$J = \frac{\partial D}{\partial t} \qquad (2.44)$$

Using Eq. 2.44 in Eq. 2.42 gives us

$$\frac{\partial v}{\partial z} = \frac{1}{c'A} \frac{\partial F}{\partial t} + \frac{e}{\epsilon c'A} I \qquad (2.45a)$$

where A is the cross-sectional area $F = TA$ and the circuit current $I = JA$. From Eq. 2.41a:

$$\frac{\partial F}{\partial z} = \rho A \frac{\partial v}{\partial t} \qquad (2.45b)$$

Equations 2.45 are now completely analogous to the equations for a transmission line excited by a current source (Eqs. 2.14), and a lumped circuit equivalent similar to Fig. 2.3b is readily derived (see Example 2.8 below). The only difference is the transformer that accounts for the factor of $e/\epsilon c'$ multiplying I in Eq. 2.45a.

Example 2.8
Calculate the $[Z]$ matrix for an acoustic transmission line segment of length ℓ described by Eqs. 2.45 and compare it to the $[Z]$ matrix for the lumped circuit in Fig. 2.9b. What should the transformer turns ratio, n, be for the two to be equivalent?

Comparing Eqs. 2.45 with Eqs. 2.14 we can make the following identifications

2 TRANSMISSION LINES AND PLANE ACOUSTIC WAVES

$$V \to -F$$
$$I \to v$$
$$L \to \rho A$$
$$C \to 1/c'A$$
$$I_s \to (e/\epsilon c'A)I$$

We can write down the first two rows of the $[Z]$ matrix from Eq. 2.20 by analogy. To get the third row we need an equation that relates the voltage V and the current I at the electrical terminals. This is obtained by integrating Eq. 2.40b from port 1 to port 2 and taking its partial time derivative.

$$I \cdot \ell = A \left[e(v_2 - v_1) + j\omega\epsilon V \right]$$

The z-matrix is written as

$$\begin{Bmatrix} F_1 \\ F_2 \\ V \end{Bmatrix} = \begin{bmatrix} -jZ_a' \cot k\ell & jZ_a' \csc k\ell & -\dfrac{je}{\omega\epsilon} \\ -jZ_a' \csc k\ell & jZ_a' \cot k\ell & -\dfrac{je}{\omega\epsilon} \\ -\dfrac{je}{\omega\epsilon} & \dfrac{je}{\omega\epsilon} & -\dfrac{j}{\omega C_0} \end{bmatrix} \begin{Bmatrix} v_1 \\ v_2 \\ I \end{Bmatrix} \qquad (2.46)$$

where $Z_a' = A\sqrt{\rho c'}$

$$v_0 = \sqrt{c'/\rho}$$
$$k = \omega/v_0$$
$$c' = c + (e^2/\epsilon)$$
$$C_0 = \epsilon A/\ell$$

We have previously calculated the $[Z]$ matrix for the lumped equivalent circuit without the transformer (see Fig. 2.3b, Example 2.5). The transformer steps up the voltage V and steps down the current I by the factor n. Consequently, the $[z]$ matrix is modified from Eq. 2.21 as follows:

$$[z] = \begin{bmatrix} -jZ_a \cot k\ell & jZ_a \csc k\ell & -\dfrac{jn}{\omega C_0} \\ -jZ_a \csc k\ell & jZ_a \cot k\ell & -\dfrac{jn}{\omega C_0} \\ -\dfrac{jn}{\omega C_0} & \dfrac{jn}{\omega C_0} & -\dfrac{jn}{\omega C_0} \end{bmatrix}$$

It is evident that equivalence is obtained if

$$\frac{n}{\omega C_0} = \frac{e}{\omega \epsilon}$$

that is, if

$$n = \frac{eC_0}{\epsilon} = \frac{eA}{\ell}$$

Example 2.9
We have seen before in Section 2.1 that the dependent and independent variables can be chosen in many different ways. Sometimes it is more convenient to use forward and reverse wave amplitudes F^+ and F^- rather than voltages and currents F and v.

$$F^+ = \frac{1}{2}(F + Z_0 v)$$

$$F^- = \frac{1}{2}(F - Z_0 v)$$
(2.47a)

Conversely,

$$F = F^+ + F^-$$

$$v = \frac{1}{Z_0}(F^+ - F^-)$$
(2.47b)

Transform the acoustic quantities in Example 2.8 to wave amplitudes.

2 TRANSMISSION LINES AND PLANE ACOUSTIC WAVES

Solution
Starting from Eqs. 2.46 and using the transformation indicated in Eqs. 2.47b, we get, after some lengthy algebra,

$$\begin{Bmatrix} F_1^- \\ F_2^+ \\ V \end{Bmatrix} = \begin{bmatrix} 0 & e^{-jk\ell} & B \\ e^{-jk\ell} & 0 & B \\ \dfrac{2B}{Z_a'} & \dfrac{2B}{Z_a'} & Z \end{bmatrix} \begin{Bmatrix} F_1^+ \\ F_2^- \\ I \end{Bmatrix} \qquad (2.48)$$

where

$$B = \frac{n}{2j\omega C_0}(1 - e^{-jk\ell})$$

$$Z = \frac{1}{j\omega C_0} + \frac{n^2}{\omega^2 C_0^2 Z_a'}(1 - e^{-jk\ell})$$

$$n = \frac{eA}{\ell}$$

2.4.2. Normal-Mode Theory

In the special case considered in Section 2.4.1, the fields are rather simple with only a few nonzero components. This made it possible to reduce the equations to a simple transmission-line form. In general, however, we shall be concerned with surface acoustic waves with both compressional and shear components having a large number of nonzero field components; the equations of motion in such cases cannot be reduced to a transmission line form. However, the generation of acoustic waves by an applied electric field can still be modeled *approximately* as a transmission line excited by a distributed current source. This model is based on the normal-mode theory, which is described in detail in Ref. 1. We will not give a rigorous derivation of this result. Instead, in this section we present the main concept using heuristic arguments and illustrate it by rederiving the exact results of Section 2.4.1 using this approach. This is the approach we will adopt in Chapter 4 to discuss the excitation of surface acoustic waves by interdigital transducers.

For a transmission line excited by a current source I per unit length, we recall from Eqs. 2.14 that

$$\frac{\partial V}{\partial z} = -L\frac{\partial I}{\partial t} \qquad (2.14a)$$

$$\frac{\partial I}{\partial z} = -C\frac{\partial V}{\partial t} + I_s \qquad (2.14b)$$

As we have seen, the solution to Eqs. 2.14 can be written as

$$V = \tilde{V}(z)\exp[j(\omega t - kz)]$$

$$I = \frac{V}{Z_0}$$

such that (see Example 2.4)

$$\frac{d\tilde{V}}{dz} = \frac{Z_0}{2} I_s e^{jkz} \qquad (2.49)$$

where $\tilde{V}$ would be constant if no source term I were present, but for nonzero I, it grows with z. Equation 2.49 thus describes the coupling of the source I to the waves in the transmission line. As long as the basic equations can be put in transmission-line form (Eqs. 2.14) we can derive an equation like Eq. 2.49 describing the growth of the wave in the presence of a source.

The difficulty with waves with more complicated fields is that we cannot cast the basic equations in transmission-line form. However, we can still derive an equation like Eq. 2.49 describing *approximately* the growth of the wave in the presence of sources. The basis for this approximate description is the reciprocity relation, discussed in detail in Ref. 1. A rigorous derivation, however, is beyond the scope of the present text. Instead, we present below a heuristic argument.

The basic idea is as follows. In general, we can calculate the fields associated with a particular wave in the *absence of sources*. In the *presence of sources* we assume that the fields are the same as those in the absence of sources, with one difference: They grow in amplitude as they propagate. The growth in the amplitude $a(z)$ is then described by

$$\frac{1}{a(z)}\frac{d}{dz}a(z) = \frac{f(z)}{2P}e^{jkz} \qquad (2.50)$$

where P = power carried by a wave of amplitude a

k = wave number of the wave being generated

f = power input per unit length from the source into a wave of amplitude a

2 TRANSMISSION LINES AND PLANE ACOUSTIC WAVES

As we mentioned earlier, a rigorous derivation of this result is beyond the scope of this text. However, we can easily see that it is quite reasonable. From the definition of f we can write

$$\frac{dP}{dz} = f(z)$$

Since $P \sim a^2$,

$$\frac{1}{2P}\frac{dP}{dz} = \frac{1}{a}\frac{da}{dz}$$

Hence

$$\frac{1}{a}\frac{da}{dz} = \frac{f(z)}{2P}$$

This is the same as Eq. 2.50 except for the factor of e^{jkz}, which is needed to reference the amplitudes of waves generated at different points to a common plane at $z = 0$.

We will now illustrate the use of Eq. 2.50 by applying it to the problem discussed in Section 2.4.1—generation of compressional waves in a piezoelectric solid by an applied electric field.

Example 2.10
Derive an equation describing the generation of a compressional wave by an electric field starting from the normal mode equation (Eq. 2.50).

Solution
Generation of compressional waves in a piezoelectric solid by an applied electric field has been discussed in Section 2.4.1. The source term is the charge that appears on the capacitor plates. The charge density $\rho(z)$ is related to the current I (Fig. 2.9a) by

$$j\omega\rho(z) = \frac{I}{A}[\delta(z) - \delta(z - \ell)] \tag{2.51}$$

In Eq. 2.50, the power input per unit length from the source $f(z)$ is given by

$$f(z) = \frac{1}{2}\tilde{\phi}^*(j\omega\rho)A \tag{2.52}$$

where $\tilde{\phi}(z)e^{j(\omega t - kz)}$ is the potential associated with the compressional wave in the absence of sources. $\tilde{\phi}(z)$ can be easily related to $\tilde{F}(z)$ by noting that in the absence of sources, $D = 0$. Hence, from Eqs. 2.40,

$$\tilde{T} = -\left(e + \frac{\epsilon c}{e}\right)\tilde{E}$$

Since $\tilde{F} = \tilde{T}A$ and $\tilde{E} = jk\tilde{\phi}$, we have

$$\tilde{F} = -jk\,\frac{\epsilon c'}{e}\,A\tilde{\phi} \qquad (2.53)$$

Identifying $\tilde{F}$ with the amplitude a in Eq. 2.51 and using Eq. 2.53, we get

$$\frac{d}{dz}\tilde{F} = \frac{\tilde{F}\tilde{\phi}^*A}{4P}\,(j\omega\rho)e^{jkz} \qquad (2.54)$$

Using Eq. 2.54 yields

$$\frac{d}{dz}\tilde{F} = \frac{Z_a'}{2}\,\frac{e}{jk\epsilon c'}\,(j\omega\rho)e^{jkz}$$

$$= \frac{eA}{2\epsilon}\,\rho e^{jkz}$$

Using Eq. 2.52 and integrating from $z = 0$ to $z = \ell$ gives us

$$\tilde{F}(\ell) = \frac{eI}{2\epsilon}\,\frac{e^{jk\ell} - 1}{j\omega}$$

assuming that $\tilde{F}(0) = 0$. $\tilde{F}(\ell)$ here is referenced to $z = 0$. Referencing it to port 2 $(z = \ell)$, we get

$$\tilde{F}_2^+ = \frac{eI}{2j\omega\epsilon}\,(1 - e^{-jk\ell}) \qquad (2.55\mathrm{a})$$

The superscript $+$ and the subscript 2 on $\tilde{F}$ indicate that the wave is traveling in the $+z$ direction and is referenced to port 2. In a similar manner we can calculate $\tilde{F}_1^-$.

$$\tilde{F}_1^- = \frac{eI}{2j\omega\epsilon}\,(1 - e^{-jk\ell}) \qquad (2.55\mathrm{b})$$

2 TRANSMISSION LINES AND PLANE ACOUSTIC WAVES

It is easy to check that this result (Eqs. 2.55) is in agreement with Eq. 2.48, derived earlier:

$$\tilde{F}_2^+ = \tilde{F}_1^- = BI$$

An interesting point to note is that we can choose any field quantity associated with a wave as its amplitude a in the normal-mode equation (Eq. 2.50). In Example 2.10 we chose $\tilde{F}$. If, instead, we choose $\tilde{\phi}$ (which is the quantity involved in the power transfer from the source, as can be seen from Eq. 2.53), the resulting equation has a particularly simple interpretation. Identifying $\tilde{\phi}$ with a in Eq. 2.50 gives us

$$\frac{d\tilde{\phi}}{dz} = \frac{Z_0}{2} j\omega\rho A \qquad (2.56a)$$

where

$$Z_0 = \frac{\tilde{\phi}^* \tilde{\phi}}{2P} \qquad (2.56b)$$

Equations 2.56 are identical to those of a *transmission line with characteristic impedance Z_0, excited by a current source $j\omega\rho A$ per unit length*. This is a rather important result that we will use in our discussion of interdigital transducers (Chapter 4) for surface acoustic waves. Such transducers produce charges at the surface of a piezoelectric solid and generate surface acoustic waves. The generation process can be modeled as a current source $j\omega\rho$ exciting the SAW transmission with a characteristic impedance $Z_0 = \tilde{\phi}^* \tilde{\phi}/2P$, where P is the power carried by a SAW with a surface potential $\tilde{\phi}$. The characteristic impedance Z_0 for various SAW substrates is discussed in Chapter 3.

Example 2.11
In Example 2.9 we derived a matrix relating F_1^-, F_2^+, V to F_1^+, F_2^-, I. Use the potentials ϕ instead of F to represent acoustic-wave amplitudes.

Solution
From Eq. 2.54,

$$\phi_1^+ = -\frac{e}{jk\epsilon c'A} F_1^+$$

$$\phi_2^+ = -\frac{e}{jk\epsilon c'A} F_2^+$$

$$\phi_1^- = \frac{e}{jk\epsilon c'A} F_1^-$$

$$\phi_2 = \frac{e}{jk\epsilon c'A} F_2^-$$

Equation 2.48 is now modified to read

$$\begin{Bmatrix} \phi_1^- \\ \phi_2^+ \\ V \end{Bmatrix} = \begin{bmatrix} 0 & e^{-jk\ell} & -r'_m \\ e^{-jk\ell} & 0 & r'_m \\ -\alpha' & \alpha' & Z \end{bmatrix} \begin{Bmatrix} \phi_1^+ \\ \phi_2^- \\ I \end{Bmatrix} \qquad (2.57)$$

where

$$r'_m = Z_0/2(1 - e^{-jk\ell})$$

$$Z_0 = K^2/\omega C_0 k\ell$$

$$K^2 = e^2/\epsilon c'$$

$$\alpha' = (1 - e^{-jk\ell}) = 2r'_m/Z_0$$

$$Z = 1/j\omega C_0 = R_0 + jX_0$$

$$R_0 = 2Z_0 \sin^2 k\ell/2$$

$$X_0 = Z_0 \sin k\ell$$

It is easy to show that Z_0 as defined here is consistent with the earlier definition in Eq. 2.56b.

In this book we use the voltage V as the independent variable, rather than I. Neglecting second-order terms such as $\alpha' r'_m/z$, we can rewrite Eq. 2.57 as

2 TRANSMISSION LINES AND PLANE ACOUSTIC WAVES

$$\begin{Bmatrix} \phi_1^- \\ \phi_2^+ \\ I \end{Bmatrix} \simeq \begin{bmatrix} 0 & e^{-jk\ell} & -\mu' \\ e^{-jk\ell} & 0 & \mu' \\ \dfrac{2\mu'}{Z_0} & -\dfrac{2\mu'}{Z_0} & Y \end{bmatrix} \begin{Bmatrix} \phi_1^+ \\ \phi_2^- \\ V \end{Bmatrix}$$

where $\mu' = r_m'/Z$

$$\simeq j\frac{K^2}{2k\ell}(1-e^{-jk\ell}) \quad [if \quad Z \simeq 1/j\omega C_0]$$

$Y = 1/Z$

The matrix elements look somewhat simpler if we reference all the amplitudes to the center of port 1 and port 2. We then have,

$$\begin{Bmatrix} \phi_1^- \\ \phi_2^+ \\ I \end{Bmatrix} = \begin{bmatrix} 0 & 1 & \mu \\ 1 & 0 & -\mu \\ -g_m & g_m & Y \end{bmatrix} \begin{Bmatrix} \phi_1^+ \\ \phi_2^- \\ V \end{Bmatrix} \qquad (2.58a)$$

where $\quad \mu = \dfrac{K^2}{2}\dfrac{\sin(k\ell/2)}{(k\ell/2)}$ \hfill (2.58b)

$g_m = 2\mu Y_0$ \hfill (2.58c)

$Y_0 = 1/Z_0 = y_0 A$ \hfill (2.58d)

$y_0 = \omega \epsilon k / K^2$ \hfill (2.58e)

$Y \simeq j\omega C_0 + G_0 + jB_0$ \hfill (2.58f)

$G_0 = (2K^2 \omega C_0 / k\ell) \sin^2(k\ell/2)$

$\quad = \mu g_m$ \hfill (2.58g)

$B_0 = (K^2 \omega C_0 / k\ell) \sin(k\ell)$ \hfill (2.58h)

Note that the ϕ's in Eq. 2.58 are different from the ϕ's in Eq. 2.57 because of the change in reference planes; however, we did not introduce new notation.

66   2 TRANSMISSION LINES AND PLANE ACOUSTIC WAVES

Example 2.12 Principle of Reciprocity
In Example 2.11 we note that $g_m = 2\mu/Z_0$. μ describes the transmitting characteristics of the transducer and g_m describes its receiving characteristic. The relationship between μ and g_m is rather fundamental and is not limited to this special case. Prove this relationship starting from the principle of reciprocity.

Solution
In Eq. 2.58a we have used wave amplitudes for ports 1 and 2, but voltages and currents for port 3. Let us also use wave amplitudes for the electric port by defining

$$V = V^+ + V^-$$

$$I = (V^+ - V^-)Y_0$$

Using V^+ and V^-, we can transform Eq. 2.58a into

$$\begin{Bmatrix} \phi_1^- \\ \phi_2^+ \\ V^- \end{Bmatrix} = \begin{bmatrix} S_{11} & S_{12} & S_{13} \\ S_{21} & S_{22} & S_{23} \\ S_{31} & S_{32} & S_{33} \end{bmatrix} \begin{Bmatrix} \phi_1^+ \\ \phi_2^- \\ V^+ \end{Bmatrix} \qquad (2.59)$$

where the elements of the scatter matrix are given by

$$S_{11} = S_{22} = \frac{\mu g_m}{Y_0 + Y}$$

$$S_{12} = S_{21} = 1 - \frac{\mu g_m}{Y_0 + Y}$$

$$S_{13} = -S_{23} = \frac{2\mu Y_0}{Y_0 + Y}$$

$$S_{31} = -S_{32} = \frac{g_m}{Y_0 + Y}$$

$$S_{33} = \frac{Y_0 - Y}{Y_0 + Y}$$

2 TRANSMISSION LINES AND PLANE ACOUSTIC WAVES

Note that $Y \simeq j\omega C_0$ while $Y_0 = \omega C_0 k\ell/K^2$ (Eq. 2.57), so that $Y \ll Y_0$ since $K^2 \ll 1$. The principle of reciprocity requires that $S_{ij} = S_{ji}^*$ for a lossless network. Hence $S_{13} = S_{31}^*$; neglecting Y relative to Y_0, this leads to

$$g_m = 2\mu Y_0 \qquad (2.60)$$

3

SURFACE ACOUSTIC WAVES IN PIEZOELECTRIC SOLIDS

A surface acoustic wave (also known as a Rayleigh wave) consists of a compressional wave and a shear wave (sometimes two shear waves) coupled together in a fixed ratio; in piezoelectric solids there is also an accompanying electrostatic wave. It decays away from the surface and is not a plane wave. In Section 3.1 we attempt to give the reader a feeling for the kinds of particle motion involved and their relationship to the electrical potential.

In Section 2.4.2 (Example 2.11 and the discussion preceding it) we used the electrostatic potential accompanying an acoustic wave to denote its amplitude. The same approach is used in Section 3.2 to develop a transmission-line model for surface acoustic waves. The voltage on the transmission line is identified with the surface electrical potential ϕ accompanying the surface wave; the particle displacements at the surface are related to the potential ϕ through constants $c_{x,y,z}$ (not to be confused with the stiffness tensor) which are listed in Table 3.1 for various substrate cuts and orientations. Table 3.2 lists the piezoelectric coupling constant K^2, equivalent dielectric constant C_s, surface-wave velocity v_0, and the characteristic admittance of the SAW transmission-line y_0. These parameters are widely used throughout the book. We should note here that the transmission-line picture for SAW is useful as long as mode conversion phenomena are not involved; the situation is similar to electromagnetic waveguides, where single mode phenomena can readily be understood in terms of transmission lines.

3 SURFACE ACOUSTIC WAVES IN PIEZOELECTRIC SOLIDS

TABLE 3.1: c_x, c_y, and c_z for Some Common Substrates

Material	Cut	Propagation Direction	c_x (Å/V)	c_y (Å/V)	c_z (Å/V)
Lithium niobate	Y	Z	0.0	$1.8j$	-1.2
	128°-rotated	X	0.1	2.0	$1.8j$
Quartz	ST	X	0.6	$-6.3j$	4.1
Lithium tantalate	77.5°-rotated Y	90° to X	0.0	-3.3	$2.9j$
Gallium arsenide	100	011	0.0	-13.1	$9.8j$

TABLE 3.2: C_s, v_0, and K^2 for Some Common Substrate Orientations

Material	Cut	Direction	C_s(pF/cm)	v_0(m/s)	y_0(mmhos)	K^2(%)
Lithium niobate	Y	Z	4.6	3488	0.21	4.6
	128°-rotated	X	5.0	3996	0.21	5.6
Quartz	ST	X	0.5	3158	0.87	0.11
Lithium tantalate	77.5°-rotated Y	90° to X	4.4	3379	0.58	1.6
Gallium arsenide	100	011	1.2	2864	3.1	0.07

In Section 3.3 a useful relationship between K^2, ϵ, v_0, and y_0 is derived for SAW, which is a slightly modified version of Eq. 2.34 derived for plane acoustic waves.

$$K^2 y_0 = 2\pi C_s v_0 \tag{3.22}$$

Finally, in Section 3.4 the factors influencing the choice of substrate cut and orientation are briefly discussed.

3.1. Introductory Description of SAW

Surface waves propagate along the surface and decay into the depth within a distance of the order of a wavelength (Fig. 3.1a). It is thus not uniform in the y direction. However, there is no variation in the transverse direction along the surface; that is, the wave is uniform in the x direction. The particles move both in the direction of wave propagation (z) and

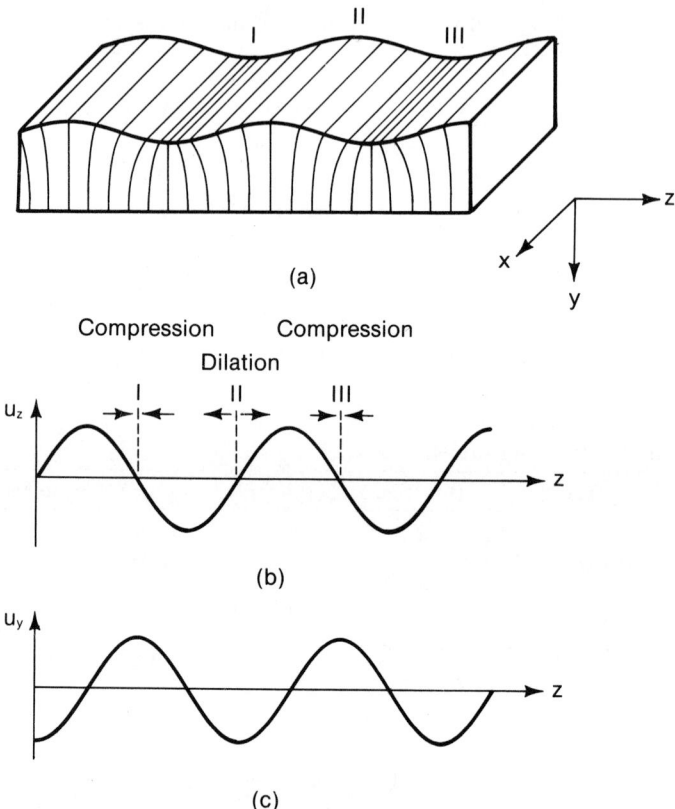

Figure 3.1 (a) Surface wave in a semi-infinite solid; (b) U_z versus z at the surface; (c) u_y versus z at the surface.

perpendicular to the depth (y) so that the wave has a mixed compressional and shear character. In some cases there is particle motion in the x direction as well.

It is easy to see why compressional and shear motion are coupled together. At a surface there is no restraining force perpendicular to the surface, so that compressing the material in the z direction automatically produces motion in the y direction. In fact, we can readily guess the phase relationship between the two particle motions u_z and u_y (Figs. 3.1b and 3.2c). The compressional motion u_z *at the surface* is shown in Fig. 3.1b. There are alternate regions of compression and dilation, marked in the figure as I, II, and III. Now, regions of compression bend down, whereas regions of dilation bulge up. The reason for this is apparent if we consider the bending of a simple rod (Fig. 3.2a). On bending, the outer surface (CD) is stretched and the inner surface (AB) is compressed (Fig. 3.2b),

3 SURFACE ACOUSTIC WAVES IN PIEZOELECTRIC SOLIDS

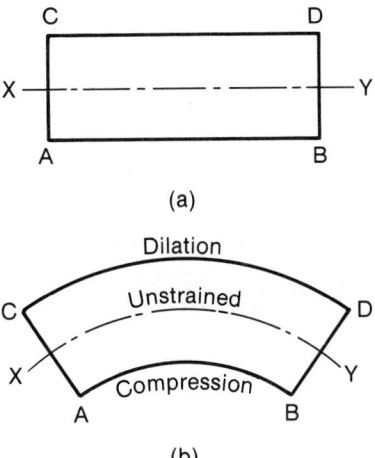

Figure 3.2. Simple illustration of how bending produces simultaneous dilation and compression: (a) straight rod; (b) bent rod with top surface (CD) dilated and bottom surface (AB) compressed. In between is a plane XY that is neither compressed nor dilated.

while there is an intermediate plane XY which is unstrained; at this plane the strain changes sign. It is clear, then, that u_y has its positive maximum in regions I and III and its negative maximum in region II, as shown in Fig. 3.1c. The shear motion u_y thus leads the compressional motion by 90° $(u_y \sim ju_z)$.

Based on Fig. 3.2, we would expect the compressional motion to *change* sign as we go into the depth. Thus at some depth in the substrate, regions I and III should show dilation while region II should show compression. This is indeed true. Figure 3.3 shows the variation of u_y and u_z into the depth along a plane perpendicular to the propagation direction (that is, at constant z). Both u_y and u_z decay as we go into the substrate; however, u_z exhibits sign reversal, as we had predicted. The detailed nature of these plots depends on the substrate material and orientation, but the qualitative nature is the same. To avoid any interference from the back surface, we usually use a substrate whose thickness is at least five wavelengths.

In SAW technology we are almost always interested in piezoelectric substrates where the acoustic fields generate electric fields. The electric field is conveniently described in terms of an electrical potential, ϕ. The electric fields E_x, E_y, and E_z are calculated from the potential by taking the negative of its derivative in the x, y, and z directions, respectively:

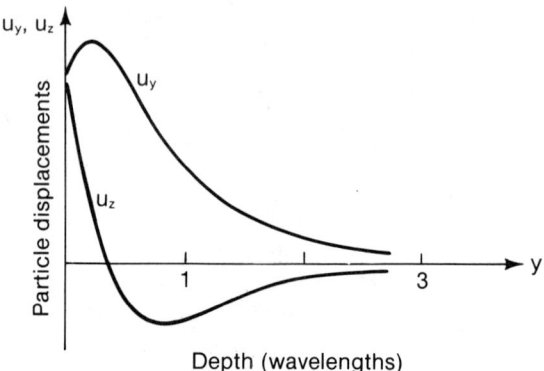

Figure 3.3 Variation of particle displacement with depth.

$$E_x = -\frac{\partial \phi}{\partial x} = 0 \tag{3.1a}$$

$$E_y = -\frac{\partial \phi}{\partial y} \tag{3.1b}$$

$$E_z = -\frac{\partial \phi}{\partial z} = jk\phi \tag{3.1c}$$

Since the wave is uniform in the x direction, $\partial/\partial x$ is zero; since it has a traveling wave character in the z direction, $\partial/\partial z$ can be replaced by $-jk$. However, it is $\partial/\partial y$ that is not simple to evaluate since the distribution in y is not a simple exponential (it can be represented as a sum of decaying exponentials). In fact, it is this feature that makes it so much more difficult to evaluate strains, stresses, and other fields for surface waves, compared to the plane uniform waves we discussed in Chapter 2.

Since a number of stress and strain components are involved here, it is difficult to predict the phase relationship of ϕ to the particle displacements. In y-cut z-propagating lithium niobate (written in abbreviated form as Y-Z LiNbO$_3$), ϕ is almost 180° out of phase with the compressional displacement u_z at the surface, and reverses sign as we go into the depth (but not at the same depth at which u_z reverses sign).

3.2. Transmission-Line Model for SAW

In SAW technology the purpose is to generate and control surface waves. This is done by means of transducers and other structures fabricated on the surface. Since all interactions with surface waves take place through the surface, we are often interested in the values of the particle displacements (u_y, u_z) or the electrical potential (ϕ) at the surface. A laser probe can be used to measure u_y at the surface; an electrostatic probe measures ϕ at the surface. So in our discussion from now on we will use the symbols u_y, u_z, and ϕ to denote the values of the corresponding quantities *at the surface*, unless specified otherwise.

The quantities ϕ, u_y, and u_z are coupled together in a particular ratio by the surface wave; they are not independent. For a particular substrate orientation, if we are given one of these, we can calculate the other two. In our discussion we will use ϕ as a measure of the amplitude of the surface wave. We could, of course, have used u_y or u_z with equal justification. However, most of the time we will be discussing the interaction of surface waves with electrical circuits; choosing ϕ as the amplitude eliminates repeated interconversion between mechanical and electrical units.

Table 3.1 lists the ratios $c_x = u_x/\phi$, $c_y = u_y/\phi$, and $c_z = u_z/\phi$ for some common substrate orientations, so that we can calculate the particle displacements if we know the electrical potential. In our discussion so far we have assumed u_x to be zero. This is true in many cases, but with some substrate orientations there is a small u_x associated with the wave. The surface wave thus consists of a compressional and two shear waves coupled together. In addition, there is an electrical wave accompanying it in piezoelectric solids. The total power carried by the wave (in the z direction) has contributions from all these waves coupled together in a fixed ratio for a given substrate.

Now we will develop a transmission-line picture for the surface wave. The surface potential ϕ of the surface wave is equated to the voltage V on the transmission line. The current I is defined so that the correct power flow is obtained; that is,

$$\phi = V \tag{3.2}$$

$$P = \frac{1}{2} VI^* = \frac{1}{2} \phi I^* \tag{3.3}$$

where P is the total power carried by the surface wave. Unlike the voltage, however, the current cannot be identified with any physical quantity associated with the surface wave. The characteristic impedance Z_0 is the ratio of the voltage V and the current I:

$$Z_0 = \frac{V}{I} = \frac{|\phi|^2}{2P} \tag{3.4}$$

where $|\phi|^2$ is the squared magnitude of ϕ equal to $\phi\phi^*$. We can now calculate the L and C parameters for the equivalent transmission line from the characteristic impedance Z_0 and the surface wave velocity v_0 (Eq. 2.6):

$$C = \frac{1}{Z_0 v_0} \qquad L = \frac{Z_0}{v_0} \qquad (3.5)$$

For a particular surface wave with a given ϕ and P, we can now calculate the characteristic impedance from Eq. 3.4 and knowing the wave velocity, the L and C parameters are obtained from Eq. 3.5. The stored energy per unit length is given by (Eq. 2.7):

$$U = \frac{1}{2} C |\phi|^2 \qquad (3.6)$$

Note that because of the way we have defined Z_0 using the total power P, Eq. 3.6 gives us the total energy stored in acoustic and electric fields, not just that in the electric field.

The characteristic impedance Z_0 as defined by Eq. 3.4, however, is not uniquely determined by the substrate orientation; it depends on the beam width W. A wider SAW beam carries more power for a given amplitude ϕ and has a lower impedance. We define a quantity z_0 which is independent of the beam width:

$$Z_0 = \frac{z_0}{W/\lambda} \qquad (3.7a)$$

We will also use the characteristic admittance Y_0, which is directly proportional to the beam width:

$$Y_0 = Z_0^{-1} = y_0 \frac{W}{\lambda} \qquad (3.7b)$$

It is important to note that the actual admittance Y_0 depends on the beam width measured in wavelengths (rather than in centimeters); the reason for this will be clear when we look at the change in field strengths with frequency. For the time being, we simply note that this is related to the fact that the SAW energy distribution in the depth (along y) changes with wavelength (see Fig. 3.3); it gets more confined at higher frequencies. Table 3.2 lists the values of y_0 and v_0 for some common substrates. The other parameters K^2 and C_s listed in the table will be discussed shortly.

3 SURFACE ACOUSTIC WAVES IN PIEZOELECTRIC SOLIDS

Example 3.1
Consider a surface wave at 100 MHz with a beam width $W = 350\ \mu$m on Y-Z lithium niobate. Calculate ϕ, u_y, and u_z if the wave carries 10 mW.

Solution

$$f = 100\ \text{MHz}$$

$$v_0 = 3500\ \text{m/s}$$

$$\lambda = 35\ \mu\text{m}$$

$$\frac{W}{\lambda} = 10$$

$$Y_0 = 2.1\ \text{mmhos} \quad \text{(Table 3.2)}$$

$$Z_0 = 476\ \Omega$$

$$|\phi|^2 = 9\ \text{V}^2 \quad (3.4)$$

$$\phi = 3\ \text{V}$$

$$u_x = c_x \phi = 0$$

$$u_y = c_y \phi = j5.5\ \text{Å} \quad \text{(Table 3.1)}$$

$$u_z = c_z \phi = -3.6\ \text{Å}$$

It is interesting to consider the change in the fields in Example 3.1 if we increase the frequency to 900 MHz but maintain the same power. Since the wavelength is now nine times as small, Z_0 will be only 50 Ω, leading to $\phi = 1$ V. Thus field quantities such as ϕ and u go down as the square root of the frequency:

$$\phi, u \propto \frac{1}{\sqrt{f}} \quad (3.8)$$

Field quantities such as S, T, E, D, and V are either the space or the time derivatives of ϕ and u. Since both space (k) and time (ω) derivatives go up linearly with frequency, we must have

$$S, T, E, D, V \propto \sqrt{f} \tag{3.9}$$

The stored energy density depends on the square of S, T, E, or D. So we expect that for a fixed total power the energy density does up with frequency:

$$U \propto f \tag{3.10}$$

This makes good sense if we remember that the wave gets more and more confined as the frequency is increased.

Now let us consider the effect of putting a thin conducting layer on the surface (Fig. 3.4). In the figure there is a little gap between the surface and the layer for clarity; however, we are considering a layer that is physically on the surface. The conducting layer shorts out the piezoelectric field (at least partly), causing a reduction in velocity. We will use K^2 to denote twice the fractional change in velocity caused by the conductor.

$$K^2 = 2 \left| \frac{\Delta v_0}{v_0} \right| \tag{3.11}$$

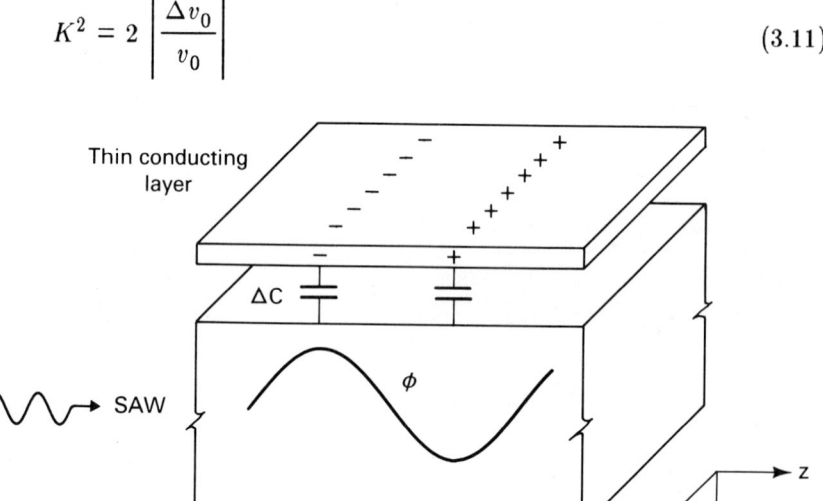

Figure 3.4 Charges induced in a thin conducting layer by a propagating surface wave.

3.3. Relation between K^2 and y_0

We will now derive a useful relationship between K^2 and y_0 (Eq. 3.22). Before going through a detailed derivation, let us see how we can get an approximate result, starting from Eq. 2.58e for plane acoustic waves.

3 SURFACE ACOUSTIC WAVES IN PIEZOELECTRIC SOLIDS

$$K^2 Y_0 = \epsilon v_0 k^2 A$$

The surface wave has a cross-sectional area $A \simeq W(1/k)$, where W is the width of the beam. Hence

$$K^2 Y_0 = \epsilon v_0 \ kW$$

so that

$$K^2 y_0 = 2\pi\epsilon v_0$$

which is Eq. 3.22. Let us now derive this relation more rigorously. The surface-wave stress fields produce an electrical potential ϕ at the surface. This potential induces charges in the conducting layer as shown in Fig. 3.4. In terms of the transmission line, it is as if we have added a capacitance ΔC (to be evaluated shortly) per unit length, causing a reduction in velocity. In fact, since $v_0 = 1/\sqrt{LC}$, for small changes in velocity we may write

$$\left|\frac{\Delta v_0}{v_0}\right| = \frac{1}{2}\frac{\Delta C}{C} \tag{3.12}$$

$$K^2 = \frac{\Delta C}{C} \tag{3.13}$$

Next we would like to find the appropriate ΔC to use. The surface charge density ρ_s induced in the sheet conductor is expected to be proportional to the potential ϕ of the surface wave. Since ρ_s is the charge per unit area, the capacitance per unit length ΔC is given by

$$\Delta C = -\frac{\rho_s W}{\phi} \tag{3.14}$$

where W is the beam width as defined earlier. The negative sign reflects the fact that negative charges accumulate in the region of positive potential (Fig. 3.4). What we have to do now is to calculate ρ_s for a given ϕ. ϕ is the potential generated at the surface by the acoustic fields in the SAW. The charge ρ_s in the conducting layer also generates electric fields in the substrate. These fields vary with z like traveling waves, that is, as e^{-jkz}. The variation with y is not obvious. We expect the fields to decay away from the charge into the substrate, but we do not know the decay rate. Let us assume that the fields produced by the charge ρ_s are described by a potential $\psi = \psi_s e^{-\alpha y} e^{-jkz}$, where α is some constant and ψ_s is its peak value at the surface.

Suppose that the substrate is isotropic with a permittivity ϵ_p. Then

$$D_y = -\epsilon_p \frac{\partial \psi}{\partial y} = \epsilon_p \alpha \psi \qquad (3.15a)$$

$$D_z = -\epsilon_p \frac{\partial \psi}{\partial z} = jk\epsilon_p \psi \qquad (3.15b)$$

Note that because of the exponential functions taking derivatives is trivial. To calculate α, we note that the divergence of D must be zero in the charge-free substrate (Maxwell's first law):

$$\frac{\partial D_y}{\partial y} + \frac{\partial D_z}{\partial z} = 0$$

Once again, taking the derivative is easy. Using Eqs. 3.15, we have

$$(\alpha^2 - k^2)\epsilon_p \psi = 0$$

that is,

$$\alpha = |k| \qquad (3.16)$$

retaining only the decaying exponential. From Eq. 3.15a we can calculate D_y:

$$D_y = \epsilon_p |k| \psi \qquad (3.17a)$$

Here we calculated the fields in the substrate. We could do a similar calculation for the medium above the charge sheet. The only changes that we need to make are to replace ϵ_p by ϵ_0 (permittivity of the medium above) and to add a minus sign to α. This is because the fields now decay in $-y$ rather than $+y$. Thus we have

$$D_y = -\epsilon_0 |k| \psi \qquad (3.17b)$$

Equation 3.17a gives us D_y below the charge and Eq. 3.17b gives us D_y above the charge. Note that D_y is discontinuous (Fig. 3.5). The discontinuity is equal to the surface charge density ρ_s (this can be proved from Maxwell's first law). Hence

$$\rho_s = (\epsilon_p + \epsilon_0) |k| \psi_s \qquad (3.18)$$

3 SURFACE ACOUSTIC WAVES IN PIEZOELECTRIC SOLIDS

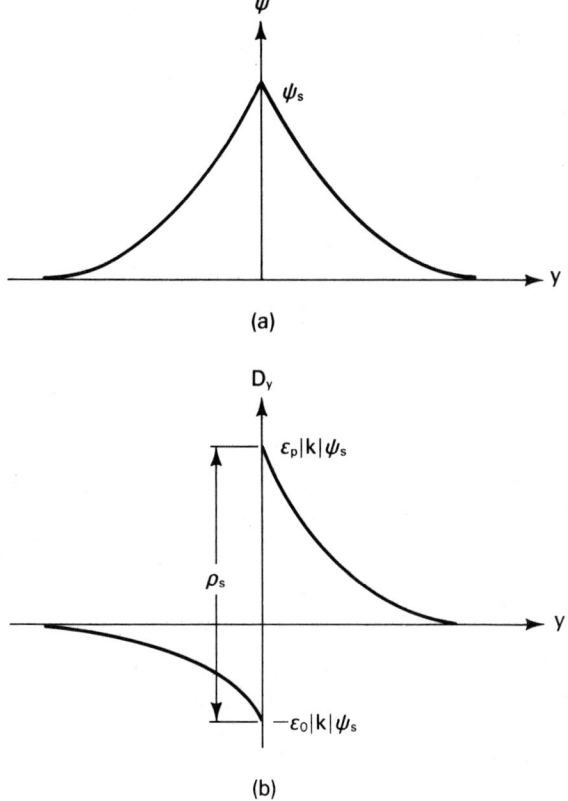

Figure 3.5 (a) ψ versus y; (b) D_y versus y.

Now we are ready to evaluate the change in capacitance per unit length (ΔC) caused by the conducting layer. Using Eq. 3.18 in Eq. 3.14, we have

$$\Delta C = -|k|(\epsilon_p + \epsilon_0) W \frac{\psi_s}{\phi}$$

Noting that $|k| = 2\pi/\lambda$, we have

$$\Delta C = 2\pi (\epsilon_p + \epsilon_0) \frac{W}{\lambda} \left[-\frac{\psi_s}{\phi} \right] \qquad (3.19)$$

ψ_s is the value of the potential produced by the charge at the surface; ϕ is the potential produced by the SAW at the surface. But these must be equal and opposite since no longitudinal electric fields (E_z) can exist in the conductor. The net potential at the surface is $\phi + \psi_s$, which implies a longitudinal electric field of $jk(\phi + \psi_s)$:

$$E_z = jk(\phi + \psi) = 0$$

$$\psi_s = -\phi \qquad (3.20)$$

Using Eq. 3.20 in Eq. 3.19 gives us

$$\Delta C = 2\pi \left(\epsilon_p + \epsilon_0\right) \frac{W}{\lambda} \qquad (3.21)$$

Using Eqs. 3.21 and 3.5 in Eq. 3.13, we get

$$K^2 y_0 = 2\pi C_s v_0 \qquad (3.22)$$

where

$$C_s = \epsilon_p + \epsilon_0$$

At this stage, let me elaborate on two points that we glossed over in our derivation when we write down D in terms of ψ (Eq. 3.15). First, we considered only an isotropic substrate. Most substrates used in SAW technology are anisotropic, so that Eq. 3.15 should be replaced by

$$D_y = \epsilon_{yy} E_y + \epsilon_{yz} E_z$$

$$D_z = \epsilon_{zy} E_y + \epsilon_{zz} E_z \qquad (3.23)$$

This poses no serious problems. We can take care of it using just the same method but with a little more algebra. The end result turns out the same, provided that we define ϵ_p by

$$\epsilon_p = (\epsilon_{yy}\epsilon_{zz} - \epsilon_{yz}^2)^{1/2} \qquad (3.24)$$

The second point is more serious. We are using a relation of the form $D = \epsilon E$ for the piezoelectric substrate rather than the more complete one $D = \epsilon E + eS$ (Eq. 2.26b). The assumption is that the acoustic fields remain undisturbed by this additional field produced by the charge. This, of course, is not precisely true, but it turns out to be a good assumption, especially if we use the permittivity ϵ at constant stress rather than that at

3 SURFACE ACOUSTIC WAVES IN PIEZOELECTRIC SOLIDS

constant strain. The last statement needs a little amplification. When we write $D = \epsilon E + eS$, we really mean the ϵ at constant strain. This is because to measure ϵ, what we have to do is look for the change in D for a small change in E keeping the *strain* (S) constant. Alternatively, D is also written as

$$D = \epsilon E + dT$$

where the ϵ is measured at constant stress (T). In strongly piezoelectric materials there is often a significant difference between the two. It is found that using the ϵ at constant stress provides a better approximation than using that at constant strain. This indicates that the charge ρ_s disturbs the strain fields of the SAW more than it disturbs the stress fields.

Example 3.2
Check to see that Eq. 3.22 is satisfied by each of the substrates in Table 3.2.

Example 3.3
How will Eq. 3.22 be modified if a conducting ground plane were present at a distance d from the surface $(d \ll \lambda)$?

Solution
In this case the fields in the substrate will be like those in a parallel-plate capacitor, so that Eq. 3.21 is replaced by

$$\Delta C = 2\pi\epsilon_0 \frac{W}{\lambda} + \epsilon_p \frac{W}{d}$$

Hence

$$K^2 y_0 = 2\pi\, C_s v_0 \frac{\epsilon_0 + \epsilon_p(\lambda/2\pi d)}{\epsilon_0 + \epsilon_p}$$

$$\simeq C_s v_0 \frac{\lambda}{d}$$

A ground plane close to the surface is often present in layered substrates such as ZnO on Si, where aluminum is deposited on Si before sputtering the ZnO.

3.4. Factors Influencing the Choice of Substrate Cut and Orientation

In Tables 3.1 and 3.2 the reader must have noticed that only specific cuts and propagation direction for a given substrate are used. We will now briefly discuss the factors that influence this choice.

The first consideration is *beam steering*. In an anisotropic solid the surface-wave velocity depends on the direction of propagation. Only those directions in which the velocity is a maximum or a minimum can be used for SAW devices. This is the main restriction on the choice of substrate orientation. For example, the SAW velocity in y-cut lithium niobate is sketched roughly in Fig. 3.6a as a function of the angle θ between the propagation direction and the z axis. The only allowable angles for SAW devices are $0°$, $\pm 25°$, and $\pm 90°$.

The reason for this is that in anisotropic media the flow of power is not necessarily perpendicular to the phase fronts. It makes an angle α with the perpendicular given by

$$\alpha = \frac{1}{v}\frac{dv}{d\theta}$$

This is known as the beam steering angle, and it makes proper design a lot more difficult. Consequently, SAW devices are always built in directions having $dv/d\theta = 0$. A simplistic view of the beam steering effect is that

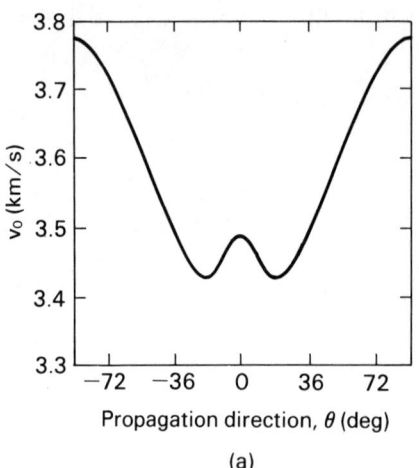
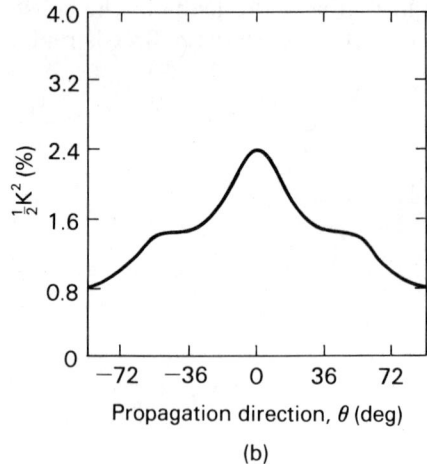

Figure 3.6 Y-cut lithium niobate: (a) surface-wave velocity; (b) coupling constant.

unless $dv/d\theta = 0$, the velocity to the left is different from the velocity to the right and the wave tends to walk off in the lower-velocity direction.

The next restriction to the choice of cut and orientation comes from the requirement for strong *piezoelectric coupling*. Because of the tensor nature of the piezoelectric constant $[e]$, the effective coupling constant K^2 depends on the cut and orientation. For example, for y-cut $LiNbO_3$ the z propagation direction has the highest K^2 and is preferred for SAW devices (Fig. 3.6b).

Apart from these principal requirements, there are requirements stemming from the particular application. For example, ST quartz is particularly suited for resonators, since the first-order temperature coefficient of velocity is zero. This leads to good temperature stability of the resonator operating frequency. Certain cuts and orientation, such as the minimal diffraction cut (MDC) of lithium tantalate, minimize diffraction losses.

4 | Part 2: SAW Device Components

INTERDIGITAL TRANSDUCERS

The *interdigital transducer* (IDT) is the most important component in SAW devices. It is used to interface between the electrical circuit and the acoustic delay line both as a transmitter (electrical → acoustic) and as a receiver (acoustic → electric) (Fig. 4.1). We need to develop an equivalent circuit for the IDT which can be used to determine its insertion loss (as a function of frequency) when acting as a transmitter (Fig. 4.3a) or as a receiver (Fig. 4.3b) in any device. Our basic aim in this chapter is to calculate the admittance of the IDT as a function of frequency. The admittance has two parts: the ordinary capacitive part, which has nothing to do with surface waves ($j2\pi f C_T$), and the acoustic admittance $Y_a(f) = G_a(f) + jB_a(f)$, which arises from the interaction of the IDT with the waves generated by it. The imaginary part of $Y_a(f)$ is called the *radiation susceptance* $[B_a(f)]$ and is equal to the Hilbert transform of the real part, which is called the *radiation conductance* $[G_a(f)]$.

In most of the literature on IDTs, $G_a(f)$ is usually computed directly. However, we find it useful to first define a transmitter response function $\mu(f)$ and a receiver response function $g_m(f)$:

$$\phi = \mu V \qquad (4.1a)$$

$$I = g_m \phi \qquad (4.1b)$$

4 INTERDIGITAL TRANSDUCERS

where ϕ is the amplitude of the SAW generated by an IDT when a voltage V is applied across its terminals and I is the current induced in a short-circuited IDT when a SAW of amplitude ϕ is incident on it from one side. The response function μ and g_m thus contain phase information, unlike G_a. This is particularly useful when interference effects are important, as, for example, in a resonant cavity with waves running in both directions; besides we feel that it lends clarity to the concepts involved.

In Section 4.1 we show that for any IDT, due to the principle of reciprocity, μ and g_m are related

$$g_m = 2\mu y_0 \frac{W}{\lambda} \tag{4.4}$$

where $y_0 \dfrac{W}{\lambda} = Y_0 = 2P/|\phi|^2$

P = power carried by a SAW of amplitude ϕ

Y_0, the characteristic admittance of the SAW transmission line, has been discussed in detail in Chapter 3. Section 4.2 is devoted to the evaluation of $\mu(f)$ for different types of IDTs. In general, it is obtained from Eq. 4.17, which requires a numerical summation. However, for IDTs with periodically spaced electrodes, the response function is the product of an element factor that is known analytically (Eq. 4.18, Fig. 4.9) and an array factor which is the Fourier transform of the tap weights (Eq. 4.16b for unapodized IDTs, Eq. 4.19 for apodized IDTs).

Section 4.3 describes how the acoustic part of the transducer admittance is calculated from the response function $\mu(f)$; it also describes how the capacitance is calculated. The radiation conductance is obtained from the response function using Eq. 4.21.

$$G_a = 2\,|\mu|^2\, y_0 \frac{W}{\lambda} \tag{4.21}$$

For apodized IDTs, Eq. 4.21 does not give the total G_a; an additional component arises from the nonuniform beam and depends on the beam profile (Eq. 4.28). The calculation of the capacitance of the IDT is discussed in Section 4.3.2. For solid-electrode IDTs, the transducer capacitance $C_T = NC_s W$, where N is the number of electrode pairs, W is the beam width, and C_s is given in Table 3.2; for split-electrode IDTs, the capacitance is 1.4 times larger (see Examples 4.17 and 4.18).

Finally, in Section 4.4, we describe a model that can be used for numerical analysis of IDTs. Unlike the analytical approach discussed earlier, this model can take into account reflection from the electrodes; it can analyze transducers whose electrodes are not periodically spaced; it also gives the radiation susceptance directly without having to compute the Hilbert transform of the radiation conductance. Usually in the literature an equivalent-circuit model (such as the Mason model, Section 2.4.2) is used to derive the transmission matrix for each electrode; transmission matrices of successive electrodes are then cascaded to provide the characteristics of the entire transducer. We have not discussed any specific equivalent-circuit models, since the transmission matrix can be obtained directly in terms of the parameters μ and g_m defined in Section 4.1.

In this chapter we do not discuss how the equivalent circuit of the IDT can be used to calculate its insertion loss. This aspect of the problem is addressed in Chapter 9.

4.1. Definition of Transmitter and Receiver Response Functions: Principle of Reciprocity

An interdigital transducer (IDT, Fig. 4.1) consists of a metallic pattern of interdigitated electrodes fabricated on the top polished surface of the piezoelectric substrate using photolithographic techniques. A complete delay line consists of two IDTs, one acting as a transmitter and the other as a receiver. The interdigital transducer (IDT) is perhaps the most important element in surface-wave devices; it provides the interface between the waves and the external circuitry. The basic quantity of interest in designing an IDT is its insertion loss as a function of frequency. This can be calculated if we know its admittance as a function of frequency.

Instead of approaching the admittance directly, we will first define two transfer functions, $\mu(f)$ and $g_m(f)$. The first describes the response of the IDT as a generator, and the second describes its response as a receiver. The admittance of the IDT and its equivalent circuit (both as a generator and as a receiver) are easily obtained from these functions. The transfer functions $\mu(f)$ and $g_m(f)$ are usually not encountered in the literature; the discussion is in terms of admittance functions directly. However, we feel that the use of transfer functions lends clarity to the concepts and brings out the phase relationship between the electrical and acoustic quantities.

The transfer functions $\mu(f)$ and $g_m(f)$ are not really independent. They are connected by the principle of reciprocity. We will now define these functions and show their inter relationship. Let us assume that transducer 1 is used as a transmitter and transducer 2 as a receiver (Fig. 4.1). A voltage V_1 applied to transducer 1 generates a SAW beam (of width W equal to the overlap between the electrodes) with an amplitude ϕ_A^+ propagating to the right. The subscript A is to indicate that the wave

4 INTERDIGITAL TRANSDUCERS

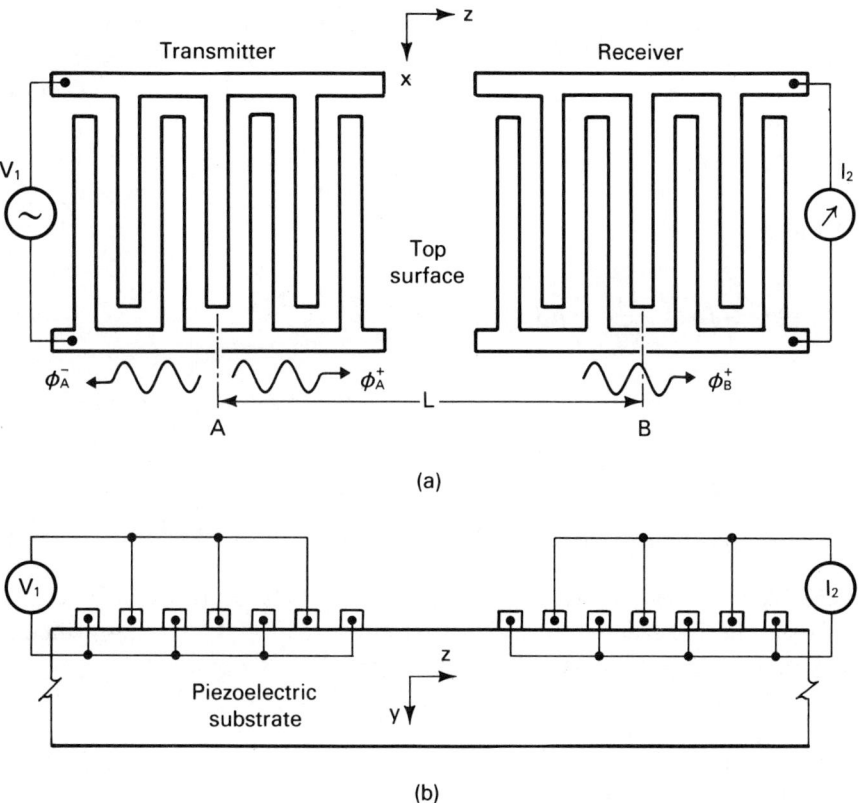

Figure 4.1 SAW delay line: (a) physical structure, top view; (b) end view.

amplitude is referenced to the center of transducer 1 (A, Fig. 4.1a). ϕ_A^+ can be related to V_1 through a function $\mu_1(f)$ describing the transmitting characteristics of transducer 1 at a frequency f:

$$\phi_A^+ = \mu_1(f) \, V_1 \tag{4.1a}$$

A SAW beam is also generated to the left, which we can neglect for our present discussion. The SAW beam generates a short-circuit current I_2 which can be related to ϕ_B^+ through a receiver response function $g_{m2}(f)$ describing the receiving characteristics of transducer 2 at frequency f:

$$I_2 = g_{m2}(f) \phi_B^+ \tag{4.1b}$$

Once again the subscript B is used to indicate that the wave amplitude is now referenced to the center of transducer 2 (B, Fig. 4.1a). The wave amplitudes at A and B are related by the propagation delay from A to B:

$$\phi_B^+ = \phi_A^+ e^{-j2\pi fT} \tag{4.1c}$$

where $T = L/v_0$, v_0 being the SAW velocity. Combining Eq. 4.1a to 4.1c, we get the transconductance transfer function $G_{12}(f)$ of the delay line from port 1 to port 2:

$$G_{12}(f) = \frac{I_2}{V_1} = \mu_1(f) g_{m2}(f) e^{-j2\pi fT} \tag{4.2}$$

Now let us reverse the roles of transducers 1 and 2 as transmitter and receiver. The transconductance transfer function from port 2 to port 1 is given by

$$G_{21}(f) = \mu_2(f) g_{m1}(f) e^{-j2\pi fT} \tag{4.3}$$

where $\mu_2(f)$ is the transmitter response function of transducer 2 and $g_{m1}(f)$ is the receiver response function of transducer 1. By the principle of reciprocity, $G_{12}(f) = G_{21}(f)$. Hence, from Eqs. 4.2 and 4.3,

$$\frac{g_{m1}(f)}{\mu_1(f)} = \frac{g_{m2}(f)}{\mu_2(f)}$$

Since transducers 1 and 2 could have totally different structures and frequency responses, this means that $g_m(f)/\mu(f)$ is independent of the nature of the transducer. In fact, for any transducer the receiver response is related to the transmitter response function by (see Example 2.12)

$$g_m(f) = 2\mu(f) Y_0 = 2\mu(f) y_0 \frac{W}{\lambda} \tag{4.4}$$

where Y_0 and y_0 are as discussed in Chapter 3. The transconductance transfer function $G_{12}(f)$ can now be written as

$$G_{12}(f) = 2\mu_1(f)\mu_2(f) Y_0 e^{-j2\pi fT} \tag{4.5}$$

which clearly reveals the reciprocity between port 1 and port 2.

Example 4.1
Consider an IDT, 350 μm wide, operating at 100 MHz on Y-Z LiNbO$_3$. It generates a SAW of amplitude 225 mV when 1 V is applied to its

4 INTERDIGITAL TRANSDUCERS

terminals. Calculate the short-circuit current that is induced in the IDT (if used as a receiver) by a 1-V surface wave.

Solution

$$v_0 \simeq 3500 \text{ m/s} \qquad \text{(Table 3.2)}$$

$$f = 100 \text{ MHz}$$

$$\lambda = 35 \ \mu\text{m}$$

$$\frac{W}{\lambda} = 10$$

$$\mu = 225 \text{ mV}/1 \text{ V} = 0.225 \qquad (4.1\text{a})$$

$$y_0 = 0.21 \text{ mmho} \qquad \text{(Table 3.2)}$$

$$g_m = 1 \text{ mmho} \qquad (4.4)$$

$$I = 1 \text{ mA} \qquad (4.1\text{b})$$

The receiving characteristics of an IDT can thus be reduced from its generating characteristics, and vice versa.

The particular conditions in Fig. 4.1 are used for illustrating the transfer function $G_{12}(f)$ and the condition of reciprocity. However, in this condition there is no power transferred from port 1 to port 2, which is not really a very useful situation. In practice, the receiver is loaded with an impedance and the transmitter is driven from a nonzero source impedance. The actual transfer function will then be different from $G_{12}(f)$. These effects can be included with a simple circuit model for the transducer, once we know its response function, $\mu(f)$ or $g_m(f)$. For example, the circuit in Fig. 4.2a can be used to determine the actual voltage V_T that appears at the IDT terminals when the voltage source has a source resistance R_s; the surface-wave amplitude is given by μV_T rather than μV. Similarly, an IDT acting as a receiver can be analyzed using the Norton's equivalent circuit shown in Fig. 4.2b.

Before proceeding, let me briefly explain how we get the different elements in Fig. 4.2 representing the admittance of the IDT. The first thing to note about the IDT is that it is basically a capacitor. In fact, if the substrate were nonpiezoelectric, it would be only that. Because we have a piezoelectric substrate, the time-varying charge distribution in the

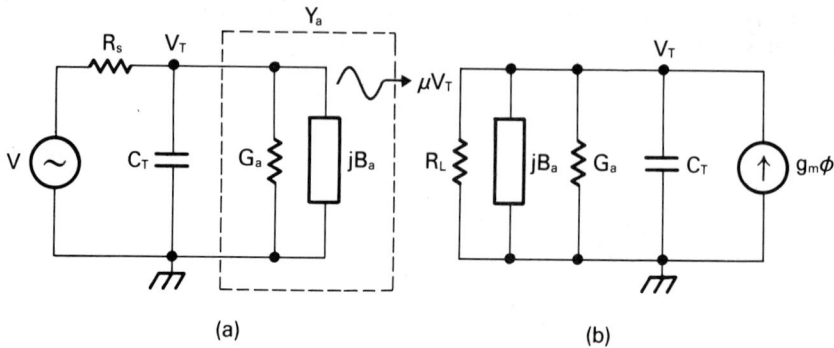

Figure 4.2 Circuit representation of an IDT: (a) transmitter; (b) receiver.

capacitor (Fig. 4.3) acts as a current generator in the SAW transmission line generating surface waves. As the surface waves propagate away from the source, they induce currents in the other electrodes and this is reflected as an admittance Y_a at the electrical terminals in parallel with the usual capacitance (C_T) (Fig. 4.2). The subscript in Y_a denotes its acoustic origin; it would be zero if the substrate were nonpiezoelectric and electrical

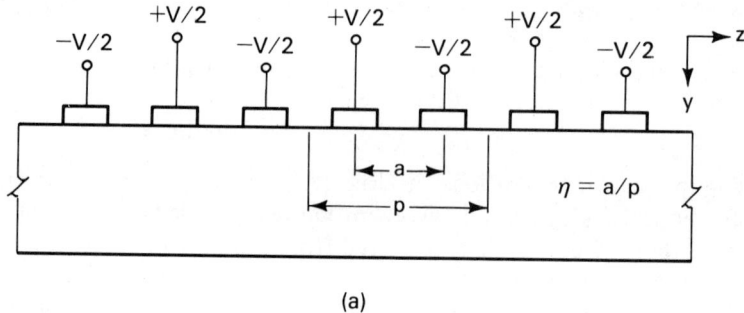

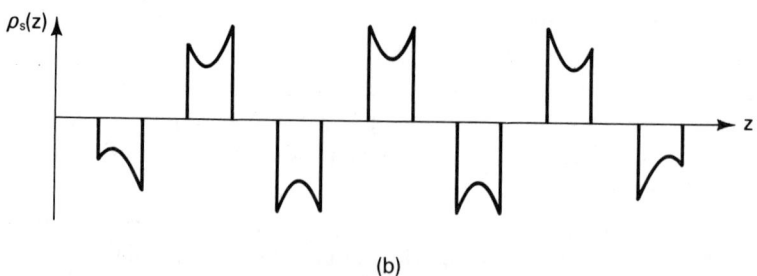

Figure 4.3 Charge distributions in an IDT: (a) electrode voltages; (b) surface charge density.

4 INTERDIGITAL TRANSDUCERS

charges did not couple to acoustic waves. As we might expect, $Y_a(f)$ is related directly to the function $\mu(f)$ describing the SAW generation characteristics of the IDT. To describe a transducer we really need to know just this function $\mu(f)$; everything else can be deduced therefrom. Our immediate objective, therefore, is to be able to calculate $\mu(f)$.

4.2. Transmitter Response Function, $\mu(f)$

4.2.1. General Theory

IDT as a distributed source. As we have mentioned, the IDT is basically a capacitor and charges appear on its electrodes when we apply a voltage to its terminals. Figure 4.3b shows the surface charge distribution $\rho_s(Z)$ in the IDT in Fig. 4.1. The charge peaks up at the edges; the reason for this can be seen as follows. The IDT can be derived from the familiar parallel-plate capacitor by rotating the plates away from each other until they are horizontal (Fig. 4.4). As we rotate the plates, the charges will pile up at the end, where the plates are closer together and the capacity is higher. The charge density ρ_s in an IDT is thus highest at the edges and decays near the center (Fig. 4.3b). Also note that the charge distribution

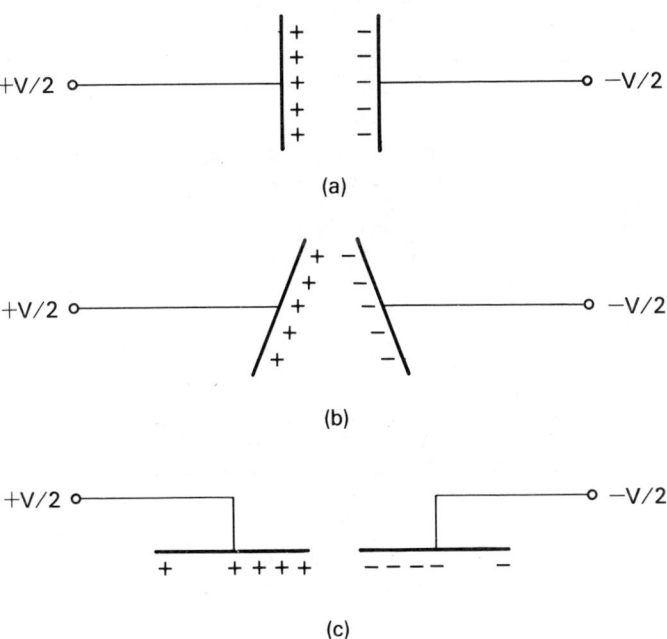

Figure 4.4. (a) Parallel-plate capacitor; (b) parallel-plate capacitor with plates sent away from each other; (c) two electrodes of an IDT.

on the end electrodes will be somewhat different from that on the central electrodes. For the present we will neglect this characteristic, which is known as the *end effect*. As we might expect, end effects are important only for short transducers, in which the end electrodes form a large percentage of the total number of electrodes.

The surface charge density $\rho_s(z)$ for a given IDT structure is determined by solving the two-dimensional Poisson's equation. This can be done on a computer, if not analytically. However, let us first see how we would go about getting the transmitter response function $\mu(f)$ if we did know the charge distribution $\rho_s(z)$.

A surface charge density of $\rho_s(z)$ means a current of $j\omega\rho_s(z)$ per unit area of the surface. This acts as a distributed current source in the SAW transmission line. The current per unit length $I(z)$ is given by

$$I(z) = j\omega\rho_s(z)W \tag{4.6}$$

Now, let us consider the generation of waves in a transmission line by a distributed current source $I(z)$. First, let us consider a small length dz' at $z = z'$ (Fig. 4.5). The current source in this length is $I(z')\,dz'$. This generates waves with amplitude (voltage) $1/2Y_0 I(z')\,dz'$ in the forward and backward directions. The factor of 2 comes from the splitting of the current in the two directions. To get the total wave generated by the distributed source, we have to add up all the waves generated at different points $z = z'$. However, wave amplitudes at different points cannot be added directly; we have to shift them all to a common reference point. Let us choose $z = 0$ as the common reference point. A wave generated at $z = z'$ shifted to $z = 0$ is

$$\frac{1}{2Y_0} I(z')\,dz'\, e^{jkz'} \quad \text{(for waves traveling to the right)}$$

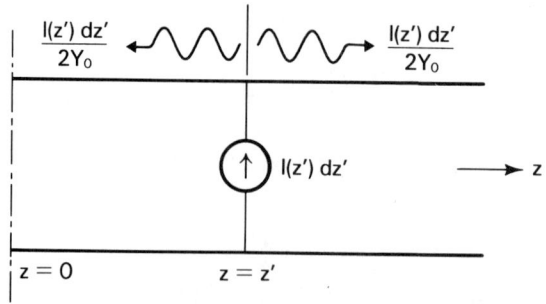

Figure 4.5 Wave generation by a point source at $z = z$.

4 INTERDIGITAL TRANSDUCERS

and

$$\frac{1}{2Y_0} I(z') \, dz' \, e^{-jkz'} \quad \text{(for waves traveling to the left)}$$

the multiplier is $e^{jkz'}$ for a shift of z' against the flow of the wave and $e^{-jkz'}$ for a shift of z' along the flow of the wave k being the wave number of the SAW.

Now we can sum up the waves generated at different points z'. The total amplitude ϕ^+ of the wave going to the right is given by

$$\phi^+ = \frac{1}{2Y_0} \int_{-\infty}^{+\infty} dz' \, I(z') e^{jkz'} \tag{4.7a}$$

and the amplitude $\phi^\pm$ of the wave going to the left is given by

$$\phi^- = \frac{1}{2Y_0} \int_{-\infty}^{+\infty} dz' \, I(z') e^{-jkz'} \tag{4.7b}$$

Replacing $I(z)$ from Eq. 4.6 in terms of $\rho_s(z)$ yields

$$\phi^+ = j \frac{K^2}{2C_s} \int_{-\infty}^{+\infty} dz' \, \rho_s(z') e^{jkz'} \tag{4.8a}$$

$$\phi^- = j \frac{K^2}{2C_s} \int_{-\infty}^{+\infty} dz' \, \rho_s(z') e^{-jkz'} \tag{4.8b}$$

Here we have used the relationship between K^2 and Y_0 (Eq. 3.22). Note that the integral in Eq. 4.8a is just the spatial Fourier transform of $\rho_s(z)$:

$$\bar{\rho}_s(k) = \int_{-\infty}^{+\infty} dz' \, \rho_s(z') e^{jkz'} \tag{4.9}$$

Thus the forward-wave amplitude is proportional to the Fourier component of ρ_s and $+k$ and the backward-wave amplitude to the component at $-k$.

$$\phi^+ = j \frac{K^2}{2C_s} \bar{\rho}_s(+k) \tag{4.10a}$$

$$\phi^- = j \frac{K^2}{2C_s} \bar{\rho}_s(-k) \tag{4.10b}$$

Equations 4.10 show that the SAW generation is frequency dependent since waves of different frequencies have different wave numbers, k.

$$k = \frac{\omega}{v_0} = \frac{2\pi f}{v_0} = \frac{2\pi}{\lambda} \tag{4.11}$$

This frequency dependence arises from the distributed nature of the generation process. The waves are generated at different points along the propagation path and they interfere differently depending on the frequency. We now have a recipe for calculating $\mu(f)$ in terms of the charge distribution:

$$\mu^{\pm}(f) = j \frac{K^2}{2C_s V} \bar{\rho}_s(\pm k) \quad k = \frac{2\pi f}{v_0} \tag{4.12}$$

where V is the transducer voltage. The charge density, of course, depends on the voltage applied so that μ is independent of voltage, as we would like it to be. We have used μ^+ and μ^- to distinguish between the SAW generation functions in the two directions. Usually, transducers are bidirectional, with $\mu^+ = \mu^-$. Real functions have symmetric Fourier transforms, so that $\bar{\rho}_s(+k) = \bar{\rho}_s(-k)$. However, it is possible to introduce phase differences between different electrodes so that the charge distribution is complex; the Fourier transform then is no longer symmetric and it is possible to build unidirectional transducers.

The spatial coordinate z and the wave number k are Fourier transform pairs in much the same way as the time t and the radian frequency ω are Fourier transform pairs. Since most of us are more familiar with time Fourier transforms, let me point out that we can deduce the frequency dependence of the SAW generation in terms of time Fourier transforms as well. Suppose that the IDT is excited with an impulse excitation at $t = 0$ (Fig. 4.6). Instantaneously at $t = 0$, a charge distribution $\rho_s(z)$ is set up. At each point a SAW is generated with an amplitude proportional to ρ_s at that point. Now let us probe the surface potential at a point along the propagation path. The SAW generated at A reaches here first, following by that at B and so on. Thus the impulse response of the system as measured at the potential probe at x will just be an inverted version of $\rho_s(z)$. In fact, the impulse response, $\phi^+(t)$, can be obtained from $\rho_s(z)$ with the substitution

$$t = -\frac{z}{v_0}$$

The frequency response of the system is the Fourier transform of the impulse response

4 INTERDIGITAL TRANSDUCERS

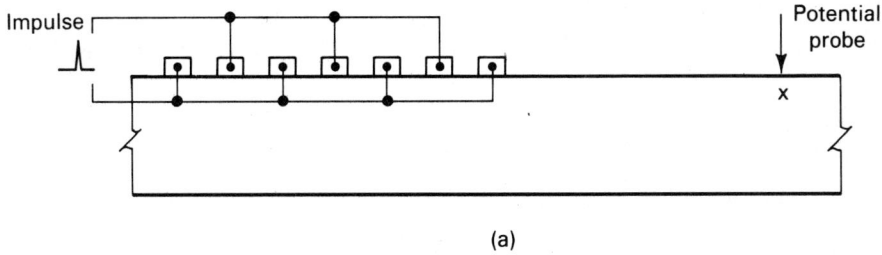

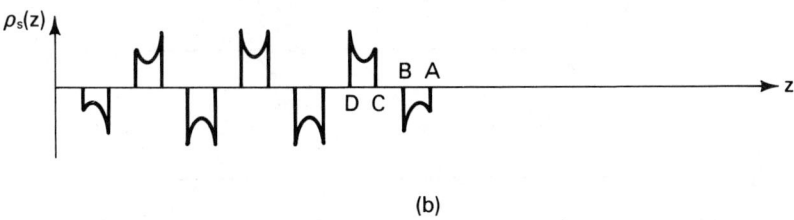

Figure 4.6 (a) Impulse excitation of an IDT; (b) instantaneous charge distribution at $t = 0$.

$$\phi^+(f) \sim \int_{-\infty}^{+\infty} dt\, \phi^+(t) e^{-j2\pi ft}$$

$$= \int_{-\infty}^{+\infty} dz\, \rho_s(z) e^{j2\pi fz/v_0}$$

which is, of course, the same result as Equation (4.8a). Thus the Fourier transform pairs z and k and t and ω differ merely by a scaling factor:

$$t = -\frac{z}{v_0}$$

$$\omega = -kv_0 \tag{4.13a}$$

This is for waves traveling in the positive z direction. For waves traveling in the negative z direction the minus signs are absent:

$$t = \frac{z}{v_0}$$

$$\omega = kv_0 \tag{4.13b}$$

4 INTERDIGITAL TRANSDUCERS

Example 4.2

Consider two oscillating line charges A and B (each 2×10^{-10} coul/cm) (Fig. 4.7) and spaced apart by 35 μm on Y-Z LiNbO$_3$. Calculate the amplitudes of the SAW generated to the left and to the right as a function of frequency for 0 to 200 MHz, if

(a) A and B have the same phase.

(b) B lags A by 90°.

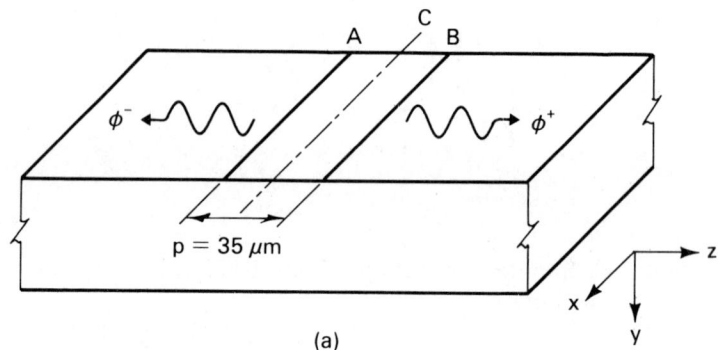

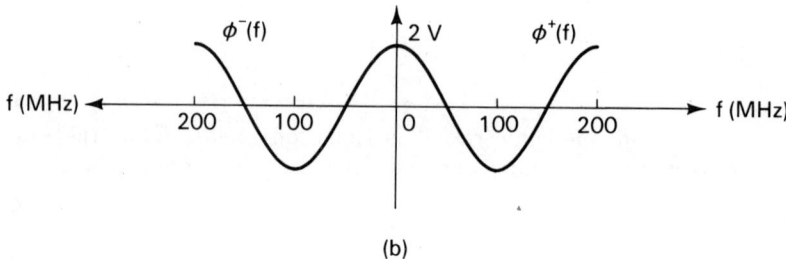

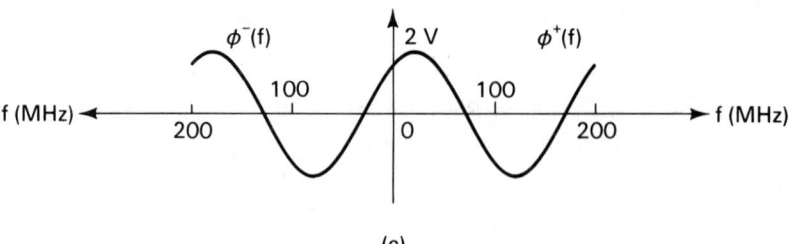

Figure 4.7 (a) SAW generation by two line charges; (b) $\phi^{\pm}(f)$ if a and b have the same phase; (c) $\phi^{\pm}(f)$ if B lags A by 90°.

4 INTERDIGITAL TRANSDUCERS

$v_0 \simeq 3500$ m/s

$y_0 = 0.21$ mmho   (Table 3.2)

$C_s = 4.6$ pF/cm

Each source is a delta function in z that generates a SAW of amplitude ϕ independent of frequency:

$$\phi = j \frac{K^2}{2C_s} \times 2 \times 10^{-1} \text{ C/cm} \qquad (4.10)$$

$$= j1 \text{ V}$$

The frequency dependence of $\phi^{\pm}(f)$ comes from the interference of the waves generated by the two charges. Using the center C as the reference plane for the wave amplitudes, we have:

Case a: $\phi^{\pm}(f) = \phi \left[e^{\pm j2\pi f p/2v_0} + e^{\pm j2\pi f p/2v_0} \right]$

Case b: $\phi^{\pm}(f) = \phi \left[e^{j\pi/4} e^{\pm j2\pi f p/2v_0} + e^{-j\pi/4} e^{\pm j2\pi f p/2v_0} \right]$

Since $2v_0/p = 200$ MHz:

Case a: $\phi^{\pm}(f) = 2j \cos(2\pi f / 2000 \text{ MHz})$ volts

Case c: $\phi^{\pm}(f) = 2j \cos \pm \dfrac{2\pi f}{200 \text{ MHz}} - \dfrac{\pi}{4}$ volts

Note that in case a the generation is bidirectional; at all frequencies, $\phi^-(f) = \phi^-(f)$. But in case b there is an asymmetry; for example, at 25 MHz there is no SAW generated to the left, while the SAW to the right is a maximum. Physically, we can see why this should be so. The SAW generated at B is 90° behind that at A. The wave from A traveling to the right is delayed by 90° when it reaches B, since at 25 MHz the distance from A to B is a quarter of a wavelength. Thus the two waves add in phase. But the wave traveling to the left from B suffers an additional delay of 90° when it reaches A and adds 180° out of phase with the wave generated to the left at A. Show that the SAW generation is bidirectional if the phase difference between A and B is 180° in this example.

Example 4.3
One of the difficulties in understanding IDTs comes from the complicated nature of the charge distribution (Fig. 4.2). Suppose that we could put a ground plane inside the substrate at a distance d below the electrodes ($d \ll p$). Each electrode would then act almost as a parallel-plate capacitor with the electric field directed along y. Calculate $\mu(f)$ for this transducer if it has only one electrode. The charge density at the electrode is constant and is equal to $C_s V/d$ if V is the voltage applied to it with respect to the ground plane.

Solution
We have to modify Eq. 4.8 slightly for this problem since Eq. 3.22 is modified when there is a nearby ground plane (see Example 3.3).

$$\phi^{\pm}(f) = j \frac{K^2}{2C_s} \int_{-a/2}^{+a/2} dz \frac{C_s V}{d} e^{-j2\pi f z/v_0} \frac{2\pi d}{\lambda} \qquad (4.8)$$

$$= jK^2 V \sin \frac{\eta \pi f}{2f_0}$$

where we have used $f_0 = v_0/2p$ and $\eta = a/p$.

$$\mu^{\pm}(f) = jK^2 \sin \frac{\eta \pi f}{2f_0}$$

At $f = 2f_0$ with $\eta = 0.5$,

$$\mu^{\pm} = jK^2$$

This problem is relevant to many transducers used with ZnO-on-Si substrates having a nearby ground plane.

Superposition principle. Various types of IDT are used to achieve different kinds of responses, each type having a different charge distribution. It appears that for any transducer we have first to determine the charge distribution, $\rho_s(z)$; we can then calculate $\mu(f)$ from Eqs. 4.12. However, as we will see shortly, all these different charge distributions are really superpositions of a certain basic charge distribution as long as the electrodes are spaced periodically. This distribution is calculated once and for all; the rest can then be deduced from it.

The principle is very simple. In any IDT we have the electrodes connected to the voltage source in a certain sequence; it is as if we have a

4 INTERDIGITAL TRANSDUCERS

number of voltage sources one at each electrode. Now, there is a well-known principle in circuit theory stating that if we have a number of voltage sources, we can look at them one at a time (shorting out the other sources) and then superpose the results. We can do the same thing here. Suppose that we have the structure in Fig. 4.8a with two electrodes connected to the voltage V and the other electrodes grounded. Intuitively, we would expect a charge distribution of the form shown. But, we may also treat this structure by superposing the two structures (Fig. 4.8b and c) shown below it, each having one electrode connected to the voltage V.

The great advantage to this approach is that the two structures being superposed are identical (except for end effects), so that we need to solve for the charge distribution in that structure just once. In fact, if we

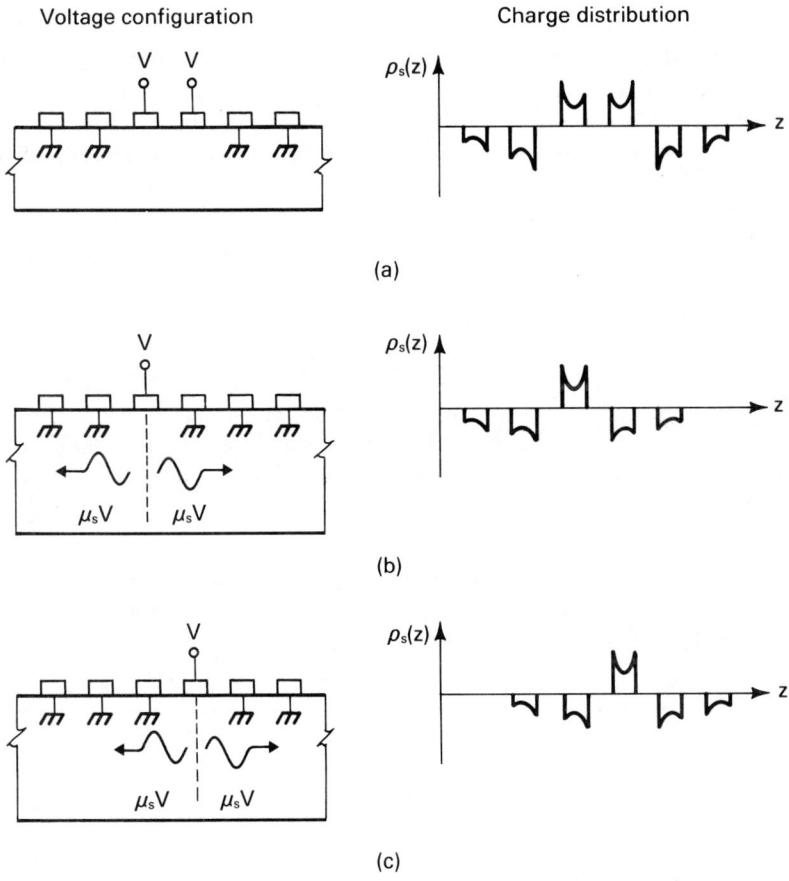

Figure 4.8 Illustration of the superposition principle. The top figure is the superposition of the two lower figures.

calculate the $\mu(f)$ for this structure [which we will call the single-tap response function $\mu_s(f)$], we can say that each electrode generates a SAW of amplitude $\mu_s(f)$ times its voltage. Then we can add up the waves generated at all the electrodes to get the total SAW generated by the transducer. Thus, if V_n is the voltage on the nth electrode located at $z = z_n$, we have

$$\phi^+(f) = \mu_s(f) \sum_n V_n e^{jkz_n} \qquad (4.15a)$$

$$\phi^-(f) = \mu_s(f) \sum_n V_n e^{-jkz_n} \qquad (4.15b)$$

Note that we have used the same function $\mu_s(f)$ for ϕ^+ and ϕ^-; that is, we have not distinguished between μ_s^+ and μ_s^-. The charge distribution in Fig. 4.8b is a real function that generates equally in either direction. A single electrode is usually bidirectional: Any unidirectional character of the overall transducer comes from the phasing of voltages between different electrodes. (There has been some very interesting work recently where unidirectionality is achieved through an interaction between transduction and reflection; we are ignoring any reflections from the electrodes.)

Another point to note is that we have taken $\mu_s(f)$ out of the summation in Eqs. 4.15. The assumption we make is that $\mu_s(f)$ is the same for each of the electrodes, which is true if we neglect end effects; otherwise, $\mu_s(f)$ for the end electrodes will be somewhat different from those in the middle. With the assumption of identical $\mu_s(f)$ for each electrode, the overall frequency response splits neatly into two factors: $\mu_s(f)$ and the Fourier transform of the voltage sequence. These are sometimes called the element factor and the array factor, by analogy with similar results in antenna theory.

Knowing $\mu_s(f)$, we can calculate the generation characteristics of an IDT with any connection sequence. If V is the voltage across the transducer terminals, then from Eqs. 4.15,

$$\mu^\pm(f) = \mu_s(f) \sum_n \frac{V_n}{V} e^{\pm jkz_n}$$

Replacing k by $2\pi f/v_0$ and introducing $f_0 = v_0/2p$, we can write this as

$$\mu^\pm(f) = \mu_s(f) H(\mp f) \qquad (4.16a)$$

where

4 INTERDIGITAL TRANSDUCERS

$$H(f) = \sum_n \frac{V_n}{V} e^{-j2\pi f z_n/v_0}$$

$$= \sum_n \frac{V_n}{V} e^{j\pi (f/f_0)(z_n/p)} \quad (4.16\text{b})$$

is the array factor. The receiver response function is written in a similar form using Eq. 4.4:

$$g_m^{\pm}(f) = g_{ms}(f) H(\mp f) \quad (4.16\text{c})$$

where

$$g_{ms}(f) = 2\mu_s(f) y_0 \frac{W}{\lambda} \quad (4.16\text{d})$$

If the electrodes are not periodic (or if we wish to account for end effects), this neat separation of element factor and array factor is not possible. We then have

$$\mu^{\pm}(f) = \sum_n \mu_{sn}(f) \frac{V_n}{V} e^{\pm j\pi (f/f_0)(z_n/p)} \quad (4.17)$$

where $\mu_{sn}(f)$ is the response function with a voltage on the nth electrode and all other electrodes grounded. This function is different for each electrode if each one is in a different environment.

Example 4.4
Consider the IDT in Fig. 4.8a with 1 V applied to two electrodes and the rest grounded. Calculate $\mu^{\pm}(f)$ if (a) the electrode voltages have the same phase; (b) the electrode to the right lags the electrode to the left by 90°.

Solution
Note that this problem is very similar to Example 4.2 except that the individual sources are not line source; they do not generate equally at all frequencies, but have the single tap response function $\mu_s(f)$.

Using Eq. 4.16b with the reference plane passing through the middle of the two electrodes:

Case (a): $H(f) = 2 \cos \dfrac{2\pi f}{200 \text{ MHz}}$

Case (b): $H(f) = 2 \cos \left[\dfrac{2\pi f}{200 \text{ MHz}} + \dfrac{\pi}{4} \right]$

Hence from Eq. 4.16a,

(a) $\mu^{\pm}(f) = 2\mu_s(f) \cos \dfrac{2\pi f}{200 \text{ MHz}}$

(b) $\mu^{\pm}(f) = 2\mu_s(f) \cos \left[\pm \dfrac{2\pi f}{200 \text{ MHz}} + \dfrac{\pi}{4} \right]$

4.2.2. Response Function of Different Transducers

Single-tap transducers. Before we go ahead and apply Eqs. 4.16 to various types of IDTs, let us see what the single tap response function looks like. To find $\mu_s(f,\eta)$ we have to solve for the charge distribution in an IDT with one electrode connected to 1 V and all others grounded (Fig. 4.8) and use this charge distribution in Eq. 4.12. This can be done analytically but the mathematics is somewhat complicated [4.3]. So let us just look at the result. Figure 4.9 shows $\mu_s(f)$ for three values of the metallization ratio η. Note that from 0 to $2f_0$ the plot looks almost like the first half-cycle of a sine wave. Actually it is not quite that, but the difference is negligible, so that we may write for $0 < f < 2f_0$,

$$\mu_s(f,\eta) \simeq \mu_s(f_0,\eta) \sin (\pi f / 2f_0) \qquad (4.18a)$$

where the peak value $\mu_s(f_0,\eta)$ depends on the metallization ratio η. The subsequent half cycles of the sine wave ($f > 2f_0$) are multiplied by P_n (cos $\eta\pi$) where P_n is the nth Legendre polynomial and n is defined as the integer part of $f/2f_0$

$$n = \text{Integer } (f/2f_0)$$

$$\mu_s(f,\eta) \simeq \mu_s(f_0,\eta) \sin \dfrac{\pi f}{2f_0} P_n (\cos \eta\pi) \qquad (4.18b)$$

Let us check if this agrees with our plot in Fig. 4.9. To do this we need to know the Legendre polynomials:

$$P_0(\cos \theta) = 1$$

4 INTERDIGITAL TRANSDUCERS

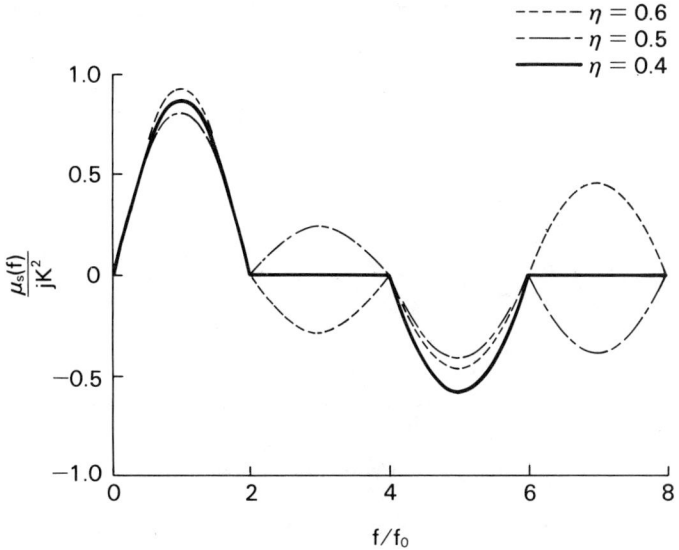

Figure 4.9 $\mu_s(f)$ versus f for $\eta = 0.4, 0.5, 0.6$.

$$P_1(\cos \theta) = \cos \theta$$

$$P_2(\cos \theta) = (3 \cos^2\theta - 1)/2$$

$$P_3(\cos \theta) = \frac{5 \cos^3 \theta - 3 \cos \theta}{2}$$

$$nP_n(\cos \theta) = (2n - 1) \cos \theta \, P_{n-1}(\cos \theta) - (n - 1) P_{n-2}(\cos \theta)$$

Thus in the first half-cycle $(0 < f < 2f_0)$, where $n = 0$, we just get back Eq. 4.18a. In the next half-cycle $(2f_0 < f < 4f_0)$ we have

$$\mu_s(f, \eta) \simeq \mu_s(f_0, \eta) \sin \frac{\pi f}{2f_0} \cos \eta\pi$$

Note that for $\eta = 0.5$, $\mu_s(f) = 0$. We can readily check the plots for the other two values of η.

The only other information we need to specify $\mu_s(f, \eta)$ completely is the value of $\mu_s(f_0, \eta)$ for different values of η. This is shown in Fig. 4.10. Note that $\mu_s(f_0)$ changes from about $0.6jK^2$ to $0.9jK^2$ as η changes from 0.25 to 0.75, which is the typical range of values in practical devices.

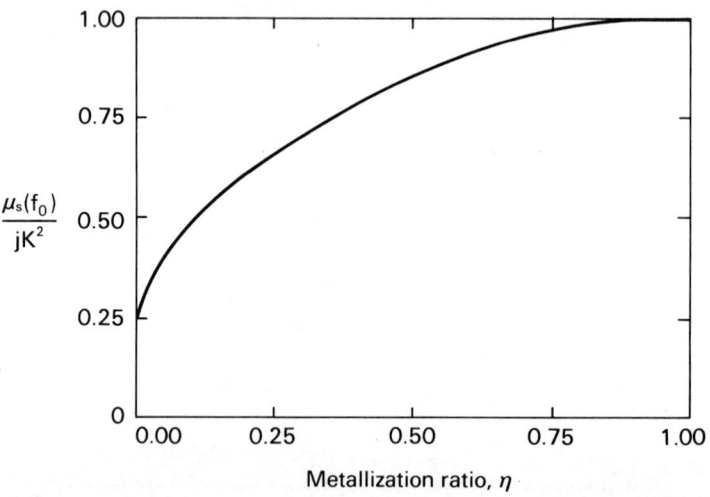

Figure 4.10 Variation of $\mu_s(f_o,\eta)$ with metallization ratio.

Example 4.5
Consider a periodic array of electrodes spaced by 17.5 μm on Y-Z LiNbO$_3$. A voltage of 1 V is applied to one electrode and all the others are grounded. Calculate the amplitude of the SAW generated at 50, 100, and 300 MHz if (a) $\eta = 0.5$; (b) $\eta = 0.25$.

Solution

$$p = 17.5 \; \mu m$$

$$v_0 = 3500 \text{ m/s}$$

$$f_0 = 100 \text{ MHz}$$

From Fig. 4.10,

$$\frac{\mu_s(f_0,n)}{jK^2} = 0.8 \quad \text{at} \quad \eta = 0.5$$

$$= 0.65 \quad \text{at} \quad \eta = 0.25$$

Using Eq. 4.18, we have

$$\frac{\mu_s(f,\eta)}{\mu_s(f_0,\eta)} = 0.7 \quad f = 50 \text{ MHz}$$

	1	$f = 100$ MHz	
	0	$f = 300$ MHz,	$\eta = 0.5$
	-0.7	$f = 300$ MHz	

$\mu_s(f, \eta)$:	f:	50	100	300	MHz
	η: 0.5	$0.026j$	$0.037j$	0	
	0.25	$0.021j$	$0.03j$	$-0.021j$	

Multiplying by the applied voltage ($= 1$ V), we get the amplitude of the generated SAW.

Unweighted transducers. Now we are ready to find the transducer response function for a few practical IDTs. Let us first take the transducer that we started with (Figs. 4.1 and 4.3). This is called the *solid-electrode* or *alternating-polarity* IDT. This transducer has a voltage sequence with alternate positive and negative voltages $V/2$ (Fig. 4.11a).

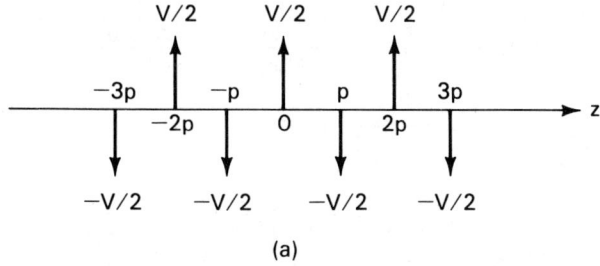

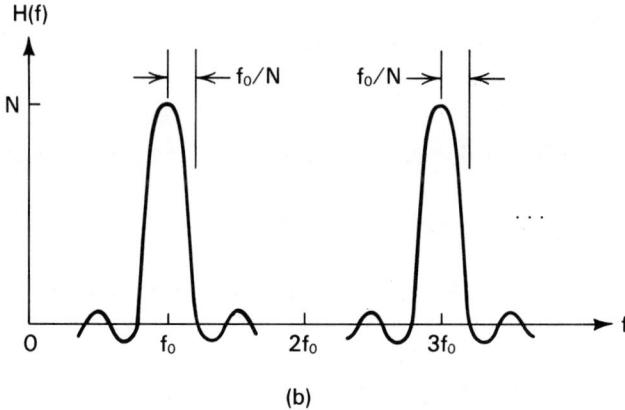

Figure 4.11 Alternating polarity IDT with N positive electrodes: (a) voltage sequence; (b) array factor.

Actually, we could have considered a voltage sequence with alternate voltages of V and 0 since adding a constant voltage to all the electrodes does not make a difference (Example 4.7). We need to evaluate the array factor for the voltage sequence in Fig. 4.11a from Eq. 4.16. This is shown in Fig. 4.11b. It consists of $[(\sin x)/x]$-shaped passbands around odd multiples of f_0. The peak value of the $(\sin x)/x$ is N, and its width as measured from the peak to the first null is f_0/N, where N is the number of positive electrodes (see Example 4.6). As we add more and more electrodes (increase N), the SAW generation gets more and more peaked. Note that the array factor is symmetric, that is, $H(f) = H(-f)$. This means the SAW generation is bidirectional, $\mu^+ = \mu^-$, so we do not distinguish between the two in the following discussion.

Now we can calculate the overall transducer response function $\mu(f)$ for the transducer by putting together the array factor $H(f)$ and the element factor $\mu_s(f, \eta)$. Note that the element factor varies with frequency rather slowly compared to the array factor, so that we may treat it as a constant around each frequency of generation. Thus the overall SAW generation function around the fundamental center frequency f_0 can be written as

$$\mu(f, \eta) \simeq \mu_s(f_0, \eta) \, N \, \frac{\sin N\pi[(f-f_0)/f_0]}{N\pi[(f-f_0)/f_0]}$$

Similarly, the SAW generation function around the third harmonic $3f_0$ can be written as

$$\mu(f, \eta) \simeq \mu_s(3f_0, \eta) \, N \, \frac{\sin N\pi[(f-3f_0)/f_0]}{N\pi[(f-3f_0)/f_0]}$$

For broadband transducers (with small N), however, the variation of the element factor with frequency should also be taken into account.

Example 4.6
Show that the array factor for the voltage sequence in Fig. 4.11a is as shown in Fig. 4.11b.

Solution
We note from Eq. 4.16 that $H(f)$ is just the Fourier transform of the voltage sequence in Fig. 4.11a if we consider it as a function of time using the transformation $t = z/v_0$. We then have a series of positive and negative impulses spaced in time by $p/v_0 = 1/2f_0$.

To start with, let us assume we have an infinite number of electrodes. If all the impulses were positive, the Fourier transform would consist of

4 INTERDIGITAL TRANSDUCERS

impulses around $n(2f_0)$. Since the impulses alternate in sign (corresponding to multiplication in time by $\cos 2\pi f_0 t$), the Fourier transform is shifted in frequency by f_0 (see the shifting theorem, Section 1.2.1). The Fourier transform of an infinite series of alternating pulses thus consists of impulses around $(2n+1)f_0$.

Actually, we have a finite number of electrodes. This means a multiplication in time by a gate function of length N/f_0 if N is the number of *positive* electrodes. In the frequency domain this corresponds to convolution with a $(\sin x)/x$ function whose width from the peak to the first null is f_0/N (see the convolution theorem, Section 1.2.2 and Example 1.7). This leads to the $H(f)$ shown in Fig. 4.11b, with $(\sin x)/x$ passbands around odd multiples of f_0.

The 4-dB bandwidth of each passband is the same as the width from the peak to the first null ($= f_0/N$). The 3-dB bandwidth is a little smaller.

$$\text{BW}_{3dB} \sim 0.9 \frac{f_0}{N}$$

Example 4.7
Calculate the array factor $H(f)$ assuming that the voltage sequence consists only of positive voltages V spaced by $2p$.

Solution
In this case we have impulses in time spaced by $2p/v_0 = 1/f_0$. So the array factor consists of $(\sin x)/x$ shapes around multiples of f_0. Note that we now have passbands around even multiples of f_0 as well (unlike Example 4.6). This, however, does not make any significant difference to the overall transducer response function since $\mu_s(f)$ is zero at the even multiples of f_0 anyway.

Example 4.8
Consider a solid electrode IDT on Y-Z LiNbO$_3$ with a fundamental center frequency of 100 MHz and a 3-dB bandwidth of 9 MHz. Calculate (a) the spacing between the electrodes; (b) the amplitude of the SAW generated at 100 MHz if 1 V is applied between its terminals.

Solution
Assume that $\eta = 0.5$.

(a) $\quad f_0 = 100 \text{ MHz} = \dfrac{v_0}{2p}$

$\quad\quad p = 17.5 \ \mu\text{m}$

(b) $\text{BW}_{3\text{dB}} = 9 \text{ MHz} = 0.9 \dfrac{f_0}{N}$

$\quad\quad N = 10$

$\quad\quad H(F) = 10 \quad \text{at } f = f_0$

Also,

$$\mu_s(f_0) = 0.037 j \quad \text{at } f = f_0, \quad \eta = 0.5 \quad \text{(Example 4.5)}$$

Hence

$$\mu = 0.37 j \quad \text{at } f = 100 \text{ MHz}$$

The amplitude of the generated SAW is 0.37 V.

Next, let us consider the double-electrode or the split-electrode IDT in which the electrodes are connected in pairs. The voltage sequence is shown in Fig. 4.12b and the corresponding array factor in Fig. 4.12c (see Example 4.9). Note that the passbands now occur around odd multiples of $f_0/2$ and the 4-dB bandwidth is $f_0/2N$, where N is the total number of *pairs* of positive electrodes.

Example 4.9
Show that the array factor for the voltage sequence in Fig. 4.12a is as shown in Fig. 4.12b.

Solution
As in Example 4.6, we note that the array factor is the Fourier transform of the voltage sequence if we convert it to a time function using $t = z/v_0$. We can think of the voltage sequence in Fig. 4.12a as the sum of the two voltage sequences shown in Fig. 4.13. We note that each of these is an array of alternating-polarity voltages very similar to that shown in Fig. 4.11. However, the voltages are spaced by $1/f_0$ rather than $1/2f_0$ and they are offset from the original by $\pm 1/4f_0$.

If we ignore the offset, we know from Example 4.6 that the Fourier transform is a series of $(\sin x/x)$ passbands around odd multiples of $f_0/2$ with 4-dB bandwidths of $f_0/2N$ (assuming N to be the number of pairs of positive electrodes in the original voltage sequence). Let us call this $H'(f)$.

4 INTERDIGITAL TRANSDUCERS

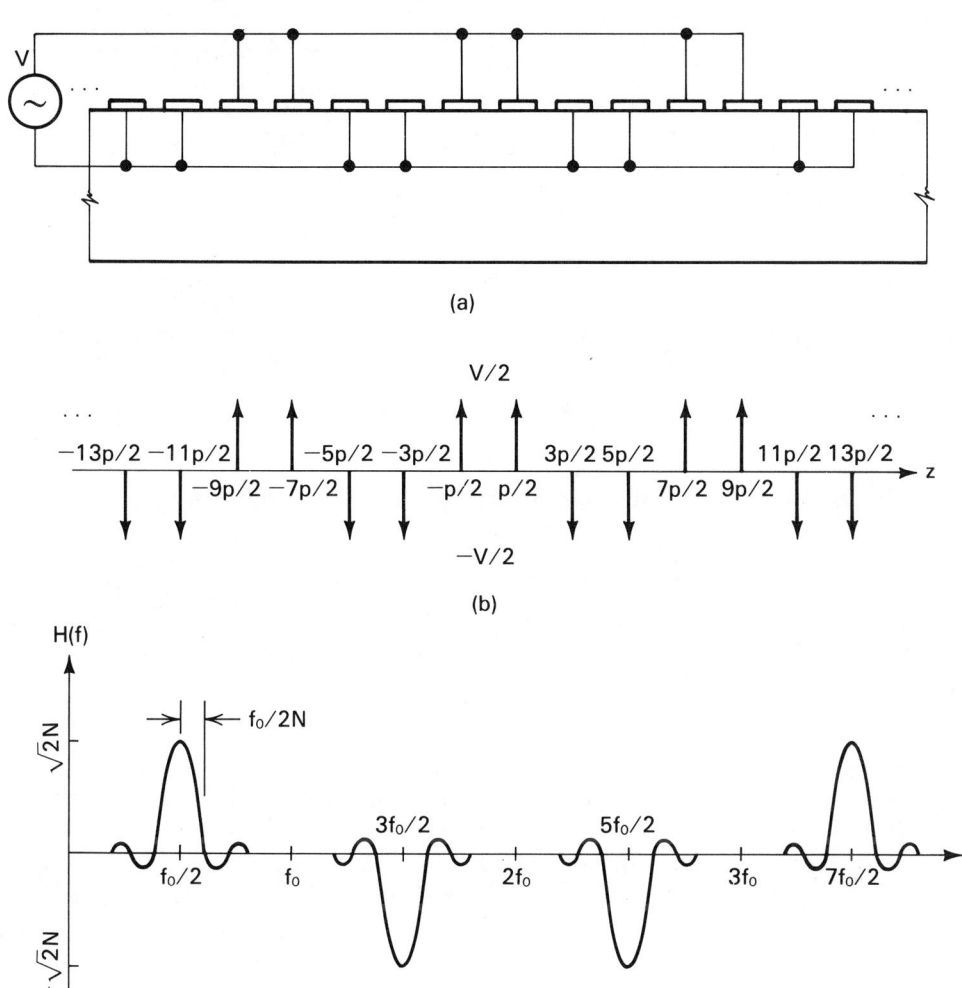

Figure 4.12 (a) Double-electrode IDT with N pairs of positive electrodes; (b) voltage sequence; (c) array factor.

The array factor $H(f)$ is then given by

$$H(f) = H'(f)\,(e^{-j2\pi f/4f_0} + e^{j2\pi f/4f_0})$$

The exponential phase factors take care of the time shifts of the sequences in Fig. 4.13 (see the shifting theorem, Section 1.2.1). Hence

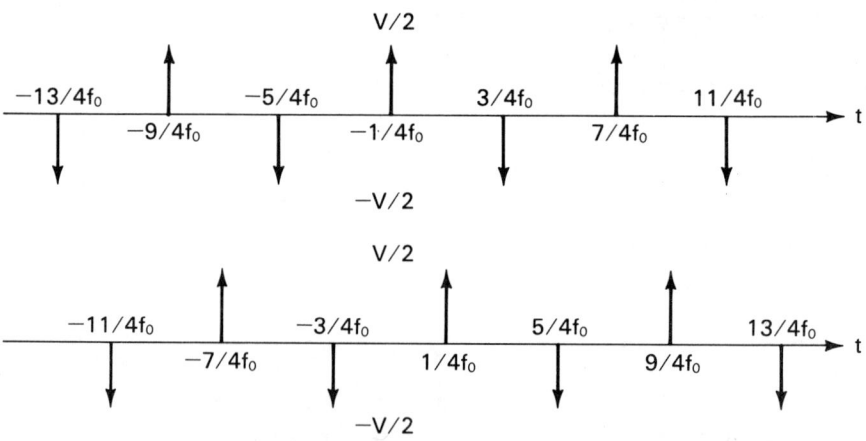

Figure 4.13 Voltage sequence in Fig. 4.12 decomposed into two parts.

$$H(f) = 2H'(f) \cos \frac{\pi f}{2f_0}$$

This is the array factor shown in Fig. 4.12b.

Example 4.10
Suppose that we wish to build a split-electrode IDT with the same fundamental center frequency and 3-dB bandwidth as the solid-electrode IDT in Example 4.8. Calculate the number of pairs of positive electrodes and the spacing between the electrodes.

Solution

Fundamental center frequency $= \dfrac{f_0}{2} = 100 \text{ MHz} = \dfrac{v_0}{4p}$

$$p = 8.75 \ \mu\text{m}$$

$$BW_{3dB} = 9 \text{ MHz} = 0.9 \dfrac{f_0}{2N}$$

$$N = 10$$

Note that the electrodes are now spaced twice as close together. The total length of the transducer, however, is the same; this is expected since the passband width is determined by the length of the transducer.

4 INTERDIGITAL TRANSDUCERS

Example 4.11
Compare the transducer response function μ at 100 MHz for the solid-and split-electrode IDTs in Examples 4.8 and 4.10.

Solution
For the solid-electrode IDT, we have seen in Example 4.8 that

$$\mu = 0.37j$$

For the split-electrode IDT at $f = 100$ MHz $(= f_0/2)$,

$$H(f) = 19\sqrt{2}$$

$$\mu_s(f) = \frac{0.027j}{\sqrt{2}}$$

so that

$$\mu = 0.37j$$

as before.

Note that the array factor for the split-electrode IDT is larger by $\sqrt{2}$, but this is exactly compensated by the element factor, which is smaller by $\sqrt{2}$. [We will see in Example 4.18 that the capacitance has increased by $\sqrt{2}$.]

Example 4.12
Compare the response function μ at the third harmonic (300 MHz) for the two IDTs in Examples 4.8 and 4.10. Assume that $\eta = 0.5$.

Solution
$H(f)$ is just as strong at 300 MHz as at 100 MHz for either IDT. For the split-electrode IDT, μ_s, too, is just as strong, so that $\mu = 0.37j$ as in Example 4.11. But for the solid-electrode IDT, μ_s at 300 MHz depends on the metallization ratio; with $\eta = 0.5$ it is zero, so that $\mu = 0$.

Before we conclude this section, let us look at the reason for using split-electrode IDTs. There is one serious problem with the solid-electrode IDT. In our analysis so far we have always neglected the reflection of the SAW off the electrodes. The electrodes present a discontinuity or a change in the characteristic impedance, causing reflections. Usually the effect is a small one. That is why we have not bothered about it so far. However, in long transducers with many electrodes, a lot of little reflections can add up to something significant. Note that at f_0 the reflections from different electrodes add up constructively. This is because the reflections off of successive electrodes have a path difference of $2p$, which at f_0 is equal to a wavelength. Thus the reflections are most severe right at the frequency where the solid-electrode IDT works. This type of transducer thus does not operate very well with many electrodes.

By contrast, the split-electrode IDT works at $f_0/2$. At this frequency, the path difference is half a wavelength and the reflections cancel each other. For this reason this type of IDT is most commonly used. The disadvantage is that the electrodes are twice as close together for the same frequency (Example 4.10), requiring better photolithographic resolution.

Example 4.13 Unidirectional Transducers
Consider the unidirectional transducer (UDT) with the voltage on each electrode lagging the preceding one by 90° (Fig. 4.14a). Calculate the array factor, $H(f)$.

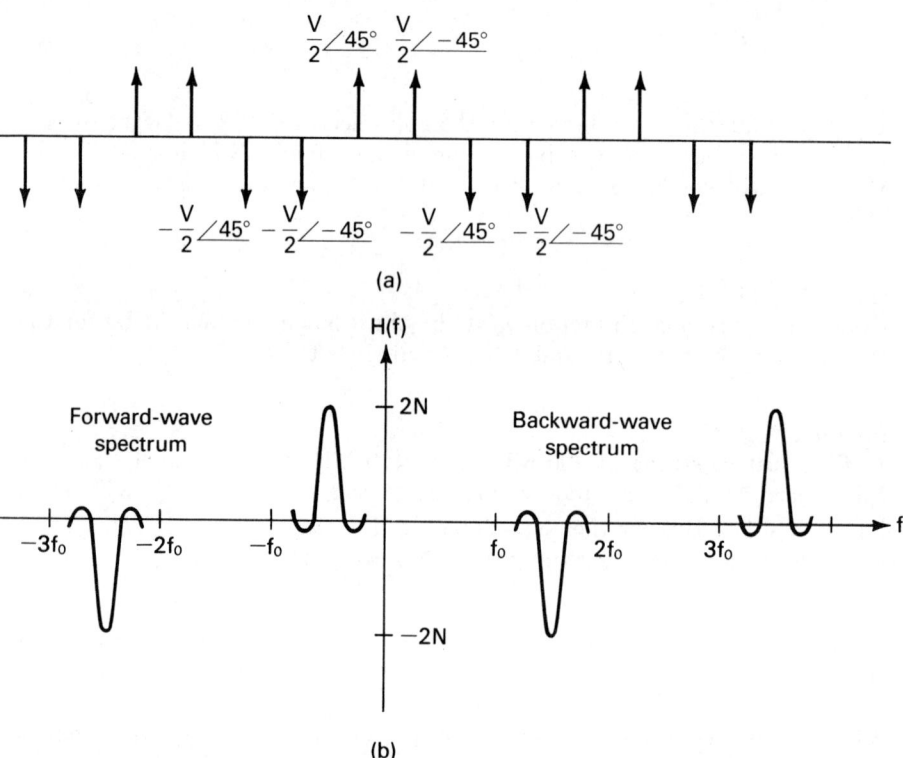

Figure 4.14 Unidirectional transducer with $2N$ positive electrodes: (a) voltage sequence; (b) array factor.

4 INTERDIGITAL TRANSDUCERS

Solution
Note that the UDT is very similar to the double-electrode IDT; the difference is that the electrode pairs are not connected to the same voltage but are phased by 90°. We can find the array factor using the same method as in Example 4.9, giving

$$H(f) = 2H'(f) \cos\left[\frac{\pi f}{2f_0} + \frac{\pi}{4}\right]$$

This is shown in Fig. 4.14b. Note that $H(f) \neq H(-f)$, so that the transducer generates unequally in the two directions.

$$\mu^{\pm}(f,\eta) = \mu_s(f,\eta) H(\mp f) \tag{4.16a}$$

It is easy to see physically why the transducer generates only in the forward direction at $f_0/2$. The waves generated at successive electrodes add constructively in this direction, but are out by phase by 180° in the backward directions (see Example 4.2).

Weighted transducers. It will be noted that all transducers that we have discussed so far give the same (sin x/x) passband shape. This is not really a good filter shape; it has a transition bandwidth equal to the filter bandwidth and the first sidelobe is only 13 dB below the main response. To synthesize arbitrary passbands, we need weighted transducers, which we will now discuss. There are two common techniques for weighting: withdrawal weighting and apodization. The former is used for narrowband filters, while the latter is used mostly for wideband filters.

Withdrawal weighted. Let us say that we wish to design an IDT with the response $R(f)$ shown in Fig. 4.15a, that is,

$$\mu_s(f) H(f) \sim R(f)$$

If we neglect the variation of $\mu_s(f)$ over the passband (this may not be acceptable for wideband filters; in that case we should correct for it), it means that we need an array factor $H(f) \sim R(f)$. Since the array factor is the Fourier transform of the voltage sequence, we require a voltage sequence that follows the impulse response $r(t)$ shown in Fig. 4.15b. However, it is difficult to vary the electrode voltages continuously from -0.5 to +0.5 in a compact fashion. So what we can do is come up with an alternative impulse response (Fig. 4.15c) that has almost the same Fourier transform around the center frequency of interest (Fig. 4.15a). The two impulse responses look very different and their Fourier transforms are far from identical. However, if we can do this right, the two can be made to

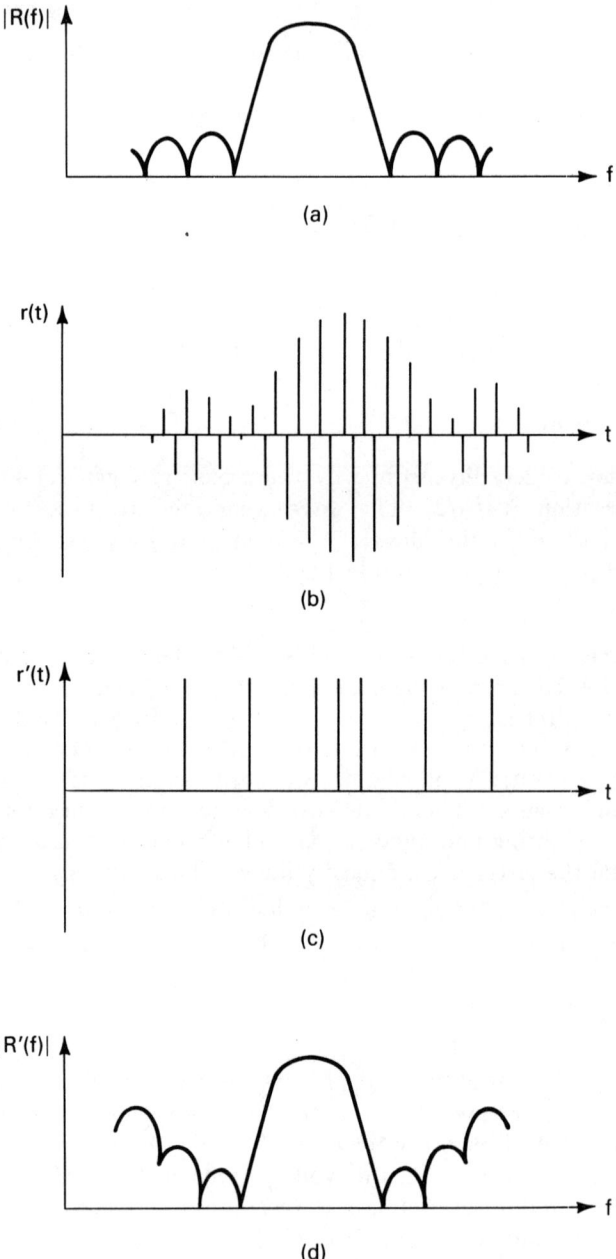

Figure 4.15 (a) Desired frequency response; (b) corresponding impulse response; (c) approximate impulse response; (d) corresponding frequency response.

4 INTERDIGITAL TRANSDUCERS

look very similar *over the narrow band of frequencies* in which we are interested. Any spurious responses outside this band can be suppressed with external matching networks.

The impulse response of Fig. 4.15c, of course, is readily implemented using either alternating-polarity or double-electrode IDTs. Note that the impulse response is nonperiodic; in some regions it is missing. In implementing the missing regions we have a choice. If we put in grounded electrodes in the missing regions, there is no problem, since the basic periodic structure of the electrodes is maintained. However, if we just leave the missing regions free of any electrodes, the periodicity of the array is repeatedly broken. So we essentially have a large number of end effects, which we neglected earlier. To account for the end effects, we have to modify $\mu_s(f, \eta)$, which is now different for each electrode depending on its environment, so that a neat separation of $\mu^{\pm}$ into two factors is not possible. We then have to use Eq. 4.17 rather than Eq. 4.16. The modification of $\mu_s(f)$ in various environments (two or three neighbors on either side describe the environment fairly well) has been computed by several authors (Refs. 4.5 and 4.6), who give tables of the tap strength in different environments relative to the periodic case. Both the amplitude and phase are affected since the charge distribution in general is no longer symmetric about the center. The transducer response is obtained by evaluating Eq. 4.17 on a computer taking into account the different values of $\mu_s(f)$ for each electrode. However, if grounded electrodes were left in place so as to maintain the periodicity, the analysis is simpler because the separation into array factor and element factor is possible. Actually, there is one good reason for withdrawing electrodes rather than putting in grounded electrodes. In high-frequency filter design, we often use alternating-polarity IDTs (rather than double-electrode IDTs) to save on the photolithographic resolution required; in that case, we have to put up with the reflections from the electrodes, and every additional electrode on the surface only makes things worse. So there is a strong incentive to withdraw electrodes if they are not generating SAWs anyway.

The design procedure for the withdrawal-weighted transducer is rather complicated (Refs. 4.7 and 4.8). Usually, what is done is to analyze all possible transducers and choose the best. If the transducer has N electrodes, the number of possible combinations is 2^N, since each electrode is either present or absent. For large N the number of combinations is extremely large; a practical approach is to break it up into groups of 8 or 10 and withdrawal weight each group separately.

Apodized. For wideband filters, an alternative weighting technique is often used. This technique, called *apodization*, varies the length of the electrodes to achieve electrode weighting (Fig. 4.16). It is a little easier to understand the operation of the apodized transducer as a receiver than as a transmitter; so let us consider this first. Suppose that we have a uniform

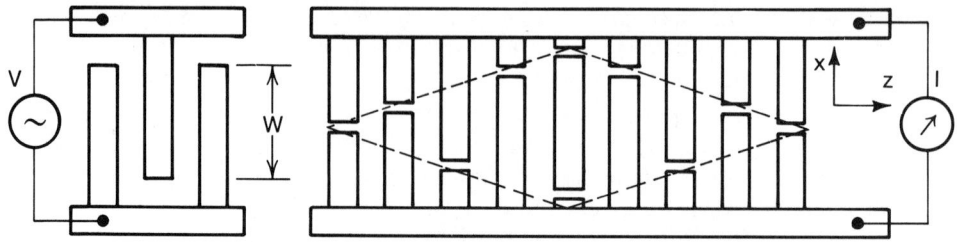

Figure 4.16 Apodized IDT used to receive uniform SAW beam of width W.

IDT generating a SAW of width W which is incident on an apodized IDT. Using the superposition principle (Section 4.2.1), we can look at one electrode at a time (with the rest grounded) and then add up the currents in all the electrodes. Suppose that we look at one electrode with the top section of length w and the lower section of length $W-w$ shown in Fig. 4.17. If w were equal to W (as in an unapodized transducer), the short-circuit current I_s would be $g_{ms}\phi/2$, where g_{ms} is given by Eq. 4.16d. In the apodized transducer, however, the currents induced in the top section (of length w) and the lower section (of length $W-w$) subtract from each other, so that

$$I_s = g_{ms}\phi \, \frac{2w - W}{2W}$$

Now we can consider the complete apodized IDT. To get the short-circuit current, we just need to add up the I_s for each of the electrodes, with appropriate phase shifts due to the propagation delay of the wave.

$$I = g_{ms} \, \phi \sum_n \frac{2w_n - W}{2W} \, e^{j\pi(f/f_0)(z_n/p)}$$

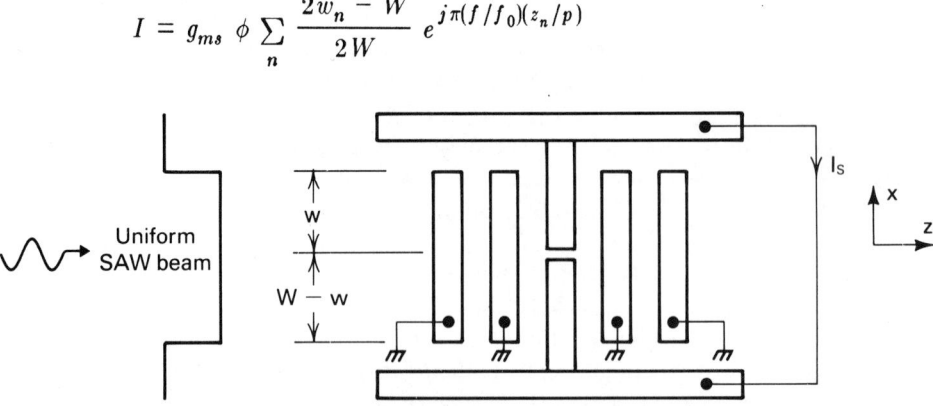

Figure 4.17 Short-circuit current induced in one electrode of an apodized IDT with other electrodes grounded.

4 INTERDIGITAL TRANSDUCERS

where the summation is to be carried out over all the electrodes, and w_n and z_n are the length and position of the nth electrode. Now, if we define the tap weight α_n of the nth electrode as

$$\alpha_n = \frac{w_n}{W} - 0.5 \tag{4.19a}$$

we have

$$g_m^{\pm}(f, \eta) = g_{ms}(f, \eta) H(\mp f) \tag{4.19b}$$

$$H(f) = \sum_n \alpha_n e^{-j\pi (f/f_0)(Z_n/p)p} \tag{4.19c}$$

This can be compared to Eq. 4.16 for unweighted transducers. We have exactly the same result, with α_n replacing V_n/V. We can now synthesize any array factor we want, since α_n is continuously variable from -1 to 1.

This illustrates the basic principle involved in apodization. An important point to note here is that if the incident beam was narrower than W, so that it intercepted only part of the apodized receiver, we would get a totally different array factor. Thus if we use one apodized IDT as transmitter and another one as receiver, the overall frequency response is *not* the product of the two individual responses. We will discuss this again, but the point to note here is that an apodized IDT works in the manner described only if excited by a uniform beam of width W.

Now, we come to the operation of the apodized IDT as a transmitter. Our previous experience with reciprocity suggests that we do not need to analyze the transmitter explicitly. The transmitter response function $\mu(f)$ must be related to $g_m(f)$ by Eq. 4.4, so that

$$\mu^{\pm}(f, \eta) = \mu_s(f, \eta) H(\mp f) \tag{4.20}$$

with $H(f)$ given by Eq. 4.19c. This is true in a certain sense but we need to reinterpret the definition of μ. Consider a single electrode in an apodized IDT (with all other electrodes grounded) being operated as a transmitter. The apodized electrode generates a SAW with amplitude $\mu_s V/2$ over the width w and $-\mu_s V/2$ over the width $W-w$. So the average amplitude is $\mu_s V[(w/W) - 0.5]$. Now if we sum the average amplitudes of the waves generated by all the electrodes with the appropriate phase shifts, we get Eq. 4.20. Note that μV gives us the average amplitude ϕ_{av} of the SAW generated by the transducer; the actual amplitude $\phi(x)$ varies along the beam. However, if we use a uniform IDT

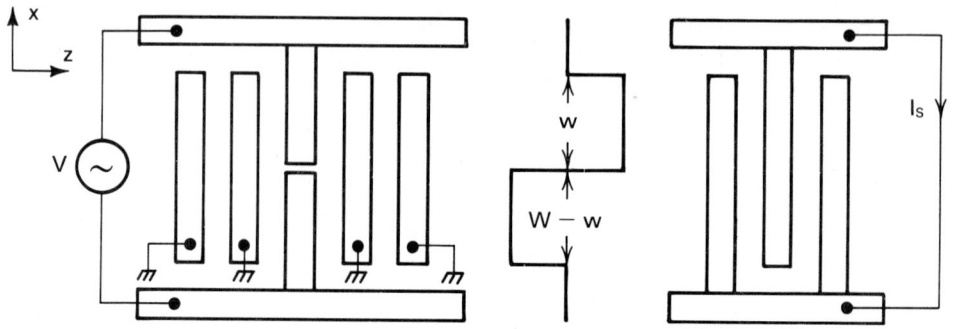

Figure 4.18 SAW generated by a single electrode in an apodized IDT with all other electrodes grounded.

to receive the beam (as in Fig. 4.18), it will average over the beam profile, so that the current I induced in it is proportional to ϕ_{av}.

Example 4.14
Consider an IDT on Y-Z LiNbO$_3$ with a triangular apodization (as in Fig. 4.16) having nine electrodes. The length w of the top section goes as 210, 105, 280, 35, 350, 35, 280, 105, 210 μm. The spacing between electrodes is 3.5 μm. Calculate (a) the fundamental center frequency; (b) the short-circuit current induced in it by a uniform SAW beam, 350 μm wide, with an amplitude of 1 V at the center frequency; (c) the average amplitude of the SAW beam generated by 1 V applied to the IDT as the center frequency. Assume that $\eta = 0.5$. The fundamental center frequency is the same as that for unapodized solid-electrode IDTs.

Solution

$$v_0 = 3500 \text{ m/s}$$
$$z_0 = 4.5 \text{ }\Omega \quad \text{(Table 3.2)}$$
$$\text{center frequency} = f_0 = \frac{v_0}{2p} = 500 \text{ MHz}$$
$$W = 350 \text{ }\mu\text{m}$$
$$\alpha_n = 0.1, -0.2, 0.3, -0.4, 0.5, -0.4, 0.3, -0.2, 0.1 \quad (4.19a)$$

At $f = f_0$:

$$H(f) = \sum_n |\alpha_n| = 2.5 \qquad (4.19c)$$

4 INTERDIGITAL TRANSDUCERS

$$\mu_s(f) = 0.037j \quad \text{(Example 4.5)}$$

$$\mu(f) = 0.093j \tag{4.20}$$

$$\lambda = 7 \ \mu m$$

$$\frac{W}{\lambda} = 50$$

$$g_m(f) = 2.1 \ j \ \text{mmhos} \tag{4.4}$$

Hence a 1-V uniform SAW beam will induce 2.1 mA of current, while 1 V applied to its terminals will generate a nonuniform SAW whose average amplitude is 0.093 V. The SAW beam will be peaked near the center of the IDT (where the generation is strongest) and will be smaller near the two edges (Fig. 4.19). The beam shape will actually change as we move off the center frequency, since the center (having more active electrodes) is more narrowband than the edges.

It will be noted that if the incident SAW beam is not uniform over the entire width of 350 μm, the frequency response would change drastically. In this example, if the incident SAW beam is only 35 μm wide around the center of the IDT, the magnitude of the current induced in all the electrodes is the same, so that the array factor has the familiar $(\sin x)/x$ shape. It is for this reason that we cannot use two apodized transducers to form a filter. This is a rather serious disadvantage, because cascading two

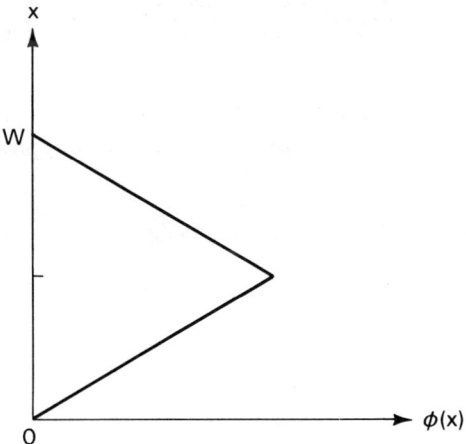

Figure 4.19 Approximate beam profile at center frequency for the apodized IDT in Example 4.14 (Fig. 4.16).

apodized IDTs would enable us to get higher out-of-band rejection. A method for cascading apodized transducers using a structure known as the multistrip coupler (MSC) will be discussed in Chapter 5. Apodized transducers coupled with MSC's are the most widely used technique for building practical wideband filters. Another disadvantage of the apodized transducer is that the varying electrode lengths lead to enhanced diffraction errors (see Section 9.4.2), so it is not used very often in making narrowband filters where a better weighting scheme (withdrawal weighting) is available.

4.3. Transducer Admittance
4.3.1. Acoustic Admittance

So far we have discussed how we can calculate the transmitter response function $\mu(f)$ and the receiver response function $g_m(f)$ for a transducer. The next thing we need is the impedance of the IDT, as seen from the electrical port. We will then have a simple circuit representation of the IDT (Fig. 4.3) that can be used to describe the interaction with external circuit elements. As we discussed before, there are two parts to the admittance of the IDT. One is $2\pi f C_T$ arising from its capacitance, which has nothing to do with acoustics, and the other is Y_a, denoting its interaction with the SAW. Let us first look at the real part of the acoustic admittance, which we will call the radiation conductance G_a. This is fairly straightforward for unapodized transducers, but for apodized transducers there is a subtle point involved.

Unapodized IDTs. If we apply a voltage V at the IDT terminals, the circuit model predicts that an amount of power $\frac{1}{2}|V|^2 G_a$ is dissipated by the IDT. This clearly must correspond to the power carried away by the SAW. We know that a SAW of amplitude $\mu^+ V$ is generated to the right and one of amplitude $\mu^- V$ to the left. Equating powers, we have

$$G_a(f) = \left[|\mu^+(f)|^2 + |\mu^-(f)|^2\right] Y_0$$

$$= 2|\mu(f)|^2 y_0 \frac{W}{\lambda} \qquad (4.21)$$

for bidirectional transducers. Here we have used the relationship between the amplitude of the SAW and the power carried by it (Section 3.2).

Apodized IDTs. With apodized IDTs the difficulty is that the generated SAW beam is nonuniform and μV gives only its average amplitude. But the power carried by a nonuniform beam cannot be obtained from the average amplitude; the power has to be averaged

4 INTERDIGITAL TRANSDUCERS

directly. Consider any nonuniform beam over the aperture O to W. The amplitude $\phi(x)$ varies along the profile, and the average amplitude ϕ_{av} is given by

$$\phi_{av} = \frac{1}{W} \int_0^W dx \, \phi(x) \qquad (4.22)$$

The power P carried by the SAW beam is given by

$$P = \frac{1}{W} \int_0^W dx \, \frac{|\phi(x)|^2}{2} y_0 \frac{W}{\lambda} \qquad (4.23)$$

Note that

$$P \neq \frac{|\phi_{av}|^2}{2} y_0 \frac{W}{\lambda}$$

Actually, there is a rather elegant way to visualize this. We can think of $\phi(x)$ as a sum of Fourier components in the x direction over the range O to W:

$$\phi(x) = \phi_{av} + \sum_n \phi_n \cos \frac{2n\pi x}{W} + \phi_n' \sin \frac{2n\pi x}{W} \qquad (4.24)$$

where ϕ_n and ϕ_n' are the amplitudes of the cosine and sine terms; for a symmetric beam, of course, the sine terms are all zero.

If we calculate $|\phi_{av}|^2/2Z_0$, it gives us the power carried by only one Fourier component, namely the uniform component. Let us call this power P_U:

$$P_U = \frac{|\phi_{av}|^2}{2} y_0 \frac{W}{\lambda} \qquad (4.25)$$

The total power is somewhat larger because of the power P_A carried by the alternating components:

$$P_A = \sum_n \left[\frac{|\phi_n|^2}{4} + \frac{|\phi_n'|^2}{4} \right] y_0 \frac{W}{\lambda} \qquad (4.26)$$

The extra factor of $\frac{1}{2}$ comes in because the average value of $\cos^2 x$ (or $\sin^2 x$) over a period is $\frac{1}{2}$. The total power $P = P_A + P_U$.

Now let us divide up the radiation conductance G_a into two conductances G_{aU} and G_{aA} in parallel; G_{aU} represents the power delivered to the uniform component and G_{aA} represents that delivered to the alternating component. G_{aU} is then given by the same expression as that for unapodized IDTs (Eq. 4.21):

$$G_{aU} = 2|\mu(f)|^2 y_0 \frac{W}{\lambda} \qquad (4.27)$$

G_{aA} depends on the shape of the beam, that is, on the Fourier component ϕ_n and ϕ_n':

$$\frac{G_{aA}}{G_{aU}} = \frac{1}{2} \sum_n \left|\frac{\phi_n}{\phi_{av}}\right|^2 + \left|\frac{\phi_n'}{\phi_{av}}\right|^2 \qquad (4.28)$$

We divided up the power in the SAW beam rather arbitrarily into alternating and uniform components. Let us look at the significance of this. A uniform IDT of width W is completely insensitive to the alternating components since they induce no total current in it; the currents induced along the IDT cancel each other out since their average is zero. So if we use a uniform IDT of width W to receive the SAW beam, the alternating components do not notice the IDT and just flow right on to the other side. The power P_A delivered to G_{aA} is undetectable and is completely lost.

Example 4.15
Consider a solid-electrode, unapodized IDT on Y-Z LiNbO$_3$, with a fundamental center frequency of 100 MHz and a 3-dB bandwidth of 9 MHz. Calculate the radiation conductance at the center frequency if the width of the IDT is 350 μm. Assume that $\eta = 0.5$.

Solution

$$\mu = 0.37j \qquad \text{(Example 4.8)}$$

$$y_0 = 0.22 \text{ mmho}$$

$$v_0 = 3500 \text{ m/s} \qquad \text{(Table 3.2)}$$

$$f = 100 \text{ MHz}$$

4 INTERDIGITAL TRANSDUCERS

$$\lambda = 35 \ \mu m$$

$$W = 350 \ \mu m$$

$$G_a = 0.61 \ \text{mmho} \tag{4.21}$$

Note that G_a is the same for the double-electrode IDT in Example 4.11 if it has the same width of 350 μm.

Example 4.16
Consider the apodized IDT in Example 4.14. Calculate G_{aU} and G_{aA} at the center frequency, assuming that the SAW beam has the triangular profile shown in Fig. 4.19.

Solution

$$\mu = 0.093 j \qquad \text{(Example 4.14)}$$

$$\frac{W}{\lambda} = 50$$

$$G_{aU} = 0.19 \ \text{mmho} \tag{4.27}$$

To calculate G_{aA} we could use Eq. 4.28, but this requires an infinite Fourier series. It is more convenient to use Eqs. 4.22 and 4.23 directly. We have

$$\frac{G_{aU} + G_{aA}}{G_{aU}} = \frac{1}{W} \int_0^W dx \ \frac{|\phi(x)|^2}{|\phi_{aV}|^2}$$

For the beam profile shown in Fig. 4.19, this gives

$$\frac{G_{aA}}{G_{aU}} = \frac{1}{3}$$

$$G_{aA} = 0.06 \ \text{mmho}$$

Note that since the beam profile changes with frequency, the ratio G_{aA}/G_{aU} also changes with frequency.

This concludes our discussion of the radiation conductance. To get the imaginary part of Y_a, which we call the radiation susceptance B_a, is

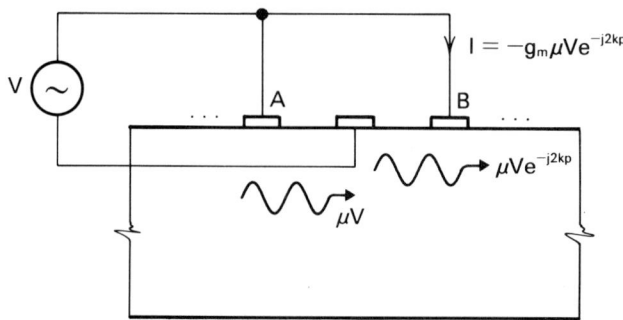

Figure 4.20 Current induced in one electrode (B) of an IDT by the wave generated at another (A).

somewhat more complicated. We will not calculate B_a rigorously, but let us take a look at what causes the susceptive part B_a to arise. Consider two electrodes A and B in an IDT (Fig. 4.20). The wave generated at A induces a current in the IDT when passing under it at B since the IDT functions equally well as a transmitter and a receiver. It is this induced current that gives rise to the acoustic admittance Y_a. In fact, we could have used this concept to calculate the radiation conductance G_a: instead, we used the power conservation concept (Eq. 4.21) because it is so much simpler. However, in calculating B_a we do not have a choice.

As each electrode generates surface waves, it also detects those generated by earlier electrodes. Let us look at the current induced in B by the wave generated at A. As we can see, this current is $(-g_m \mu e^{-j2kp})V$. At the center frequency, p is an exact multiple of a wavelength, so that $e^{-j2kp} = 1$. The induced current is then $-g_m \mu V$, which is in phase with the voltage. We then expect that at the center frequency, B_a should be zero. However, if we move to a slightly higher frequency, the current lags the voltage (because p is a little more than a wavelength, causing a little extra delay). This produces a negative B_a (inductive). Conversely, at a frequency below the center frequency we get a positive B_a (capacitive) (Fig. 4.21). Here we have considered just two electrodes. To calculate B_a exactly, we need to consider all pairs of electrodes and sum up the currents, which is fairly simple in principle. A transmission-line model does this automatically (Section 4.4). Actually, there is a general principle for causal systems that requires $B_a(f)$ to be the Hilbert transform of $G_a(f)$. Knowing $G_a(f)$, we could take its Hilbert transform to get $B_a(f)$.

$$B_a(f) = \frac{1}{\pi} \int_{-\infty}^{+\infty} du \, \frac{G_a(u)}{u - f} \qquad (4.29)$$

4 INTERDIGITAL TRANSDUCERS

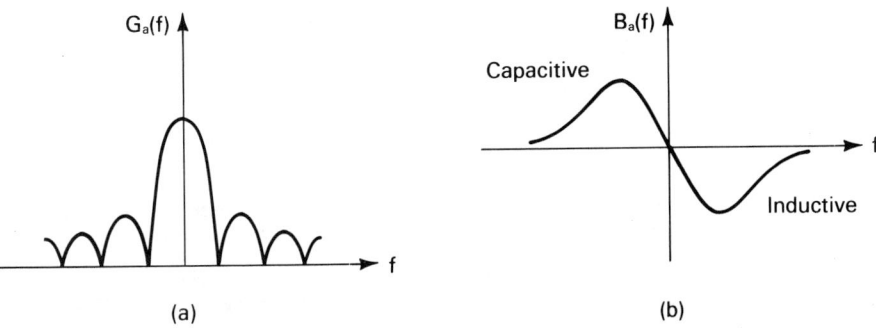

Figure 4.21 $Y_a(f)$ for an IDT around the center frequency f_o: (a) $G_a(f)$ versus f; (b) $B_a(f)$ versus f.

For an IDT with a $(\sin x)/x$ frequency dependence of μ, $G_a(f)$ has a $(\sin^2 x)/x^2$ dependence $[x = N\pi(f - f_0)/f_0]$. This leads to a $B_a(f)$ of the form

$$B_a(f) = G_a(f_0) \frac{\sin 2x - 2x}{2x^2}$$

This is shown in Fig. 4.21. Note that $B_a(f = f_0) = 0$, as we reasoned earlier. This means that for calculations at the center frequency we do not need $B_a(f)$. There is a rather general principle here (which appears in many different contexts) about the Hilbert transform and why causality requires that the real and imaginary parts of $Y_a(f)$ be related by the Hilbert transform.

The Hilbert transform (Eq. 4.29) can be interpreted as a convolution in the frequency domain of the function $1/\pi f$ with $G_a(f)$. Now, convolution in the frequency domain corresponds to multiplication in the time domain, so it is worth asking what the function $1/\pi f$ corresponds to in the time domain. If you work it out (it is easier to work back from the result), it turns out to be $-j \, \text{sgn}(t)$, where

$$\text{sgn}(t) = \begin{cases} 1 & t > 0 \\ -1 & t < 0 \end{cases}$$

So the Fourier transform of $B_a(f)$ is equal to $-j \, \text{sgn}(t) \, g_a(t)$, where $g_a(t)$ is the Fourier transform of $G_a(f)$. Hence

$$F^{-1}\{Y_a(f)\} = y_a(t) = g_a(t) + jb_a(t)$$
$$= g_a(t) + \text{sgn}(t) \, g_a(t)$$

It is apparent that $y_a(t)$ is necessarily causal, as all negative time components are canceled by $[1 + \text{sgn}(t)]$. Of course, the Fourier transform of $Y_a(f)$ must be causal, since it is the impulse response and there can be no response before the excitation.

4.3.2. Capacitance

Now that we know how to calculate the acoustic part of the admittance, let us turn to the capacitance. Basically, this involves a two-dimensional electrostatic field problem (three-dimensional for apodized IDTs). The results, of course, depend on the particular type of connection used and the metallization ratio, η. The interesting thing, though, is that the capacity does not depend on the absolute value of the electrode spacing p or the electrode width a, but on the metallization ratio of $\eta = a/p$. If we think of the IDT as a parallel-plate capacitor with the plates rotated away from each other (Fig. 4.4), we can see that while the plate area depends on a, the plate separation depends on p; this explains why doubling both a and p does not change the capacitance.

A useful representation of the IDT is in terms of capacitors $C_1, C_2, \ldots, C_n$ connecting each electrode to its 1st, 2nd, ..., nth neighbor (Fig. 4.22). As we may expect, the C's depend on the metallization ratio. However, at $\eta = 0.5$ we have a remarkably simple result (Ref. 4.1):

$$C_n = \frac{4}{\pi} C_s W \frac{1}{4n^2 - 1} \tag{4.30}$$

This is not meant to be obvious, but we will not prove it here. With this result it is quite straightforward to calculate the capacitance of any IDT

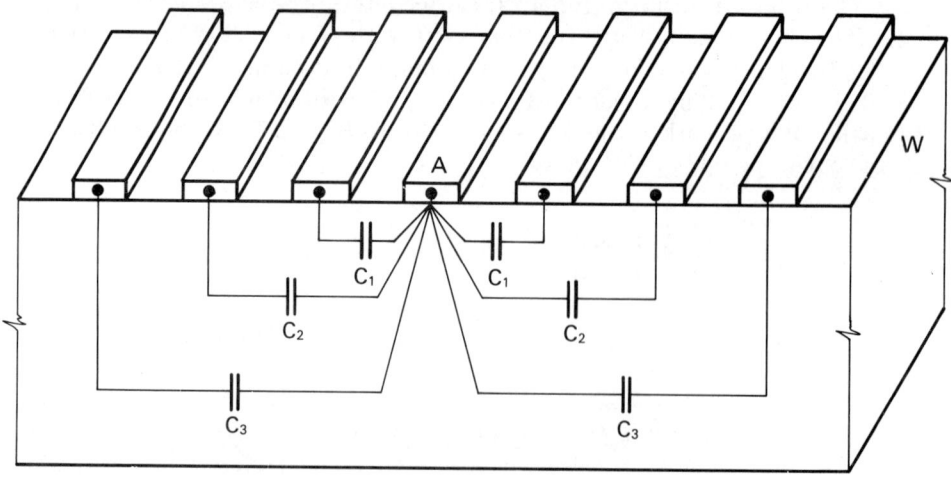

Figure 4.22 Capacitors connecting electrode A to its neighbors.

4 INTERDIGITAL TRANSDUCERS

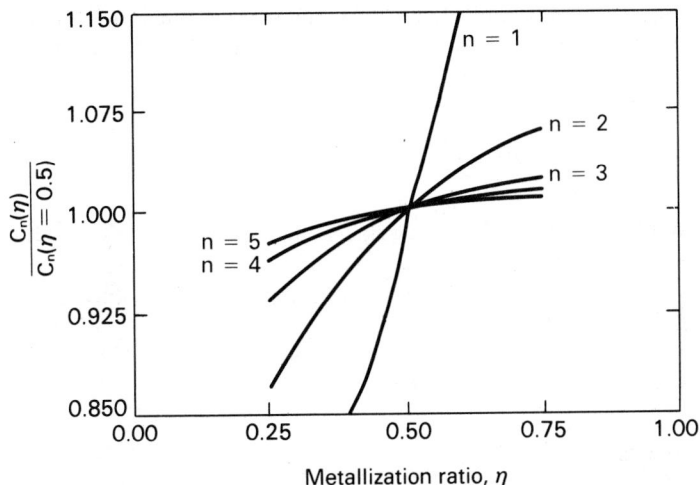

Figure 4.23 Variation of C_n's with metallization ratio.

(see the following examples), provided that the electrodes are periodically spaced. With nonperiodic electrodes, however, Eq. 4.30 is not valid.

The C_n's given by Eq. 4.30 are for a metallization ratio of 50%. As we might expect, the C_n's depend on η. This is particularly true for the nearer neighbors; C_1 changes more than C_2, which, in turn, changes more than C_3, and so on. In fact, as Fig. 4.23 shows, the C_n's for $n > 3$ hardly change with η. A good approximation for C_1 and C_2 in the range $0.25 < \eta < 0.75$ (which is the practical range) is given by

$$C_1(\eta) = C_1(\eta = 0.5) \exp[1.75(\eta - 0.5)] \tag{4.31a}$$

$$C_2(\eta) = C_2(\eta = 0.5) (\eta/0.5)^{0.18} \tag{4.31b}$$

These equations have been obtained by curve fitting to the actual curves in the range mentioned.

Example 4.17 Solid-Electrode IDT
Calculate the capacitive admittance of the solid-electrode IDT at the center frequency given in Examples 4.8 and 4.15.

Solution
Let us look at one positive electrode in the IDT. It is connected through capacitors C_1, C_3, C_5, ... to electrodes of the opposite polarity. The even capacitors connect it to an electrode with the same voltage and can be

disregarded (Figs. 4.2 and 4.22). So the capacitance for each positive electrode is the sum of all the odd capacitors. If N is the total number of positive electrodes,

$$C_T = N \sum_{\text{odd } n} 2C_n = NC_S W$$

(the factor of 2 accounts for the two capacitors, one on each side), where C_T is the transducer capacitance.

$$f = 100 \text{ MHz}$$

$$N = 10 \quad \text{(Examples 4.8, 4.15)}$$

$$W = 350 \text{ } \mu\text{m}$$

$$C_s = 4.6 \text{ pF/cm} \quad \text{(Table 3.2)}$$

Hence

$$C_T = 1.61 \text{ pF}$$

$$2\pi f C_T = 1.01 \text{ mmhos}$$

Example 4.18 Split-Electrode IDT
Calculate the capacitive admittance of the double-electrode IDT at the center frequency given in Example 4.11. $W = 350$ μm.

Solution
From Fig. 4.12a we see that for the double-electrode IDT,

$$C_T = 2N(C_1 + 2C_2 + C_3 + C_5 + 2C_6 + C_7 + \cdots)$$

$$= 1.4 N C_s W$$

Hence

$$C_T = 2.25 \text{ pF}$$

$$2\pi f C_T = 1.4 \text{ mmhos}$$

4 INTERDIGITAL TRANSDUCERS

Note that the capacitance of the split-electrode IDT is 1.4 times higher, although the radiation conductance is the same as that of the solid-electrode IDT.

Example 4.19 Apodized IDT
Consider the apodized IDT in Examples 4.14 and 4.16. Calculate its capacitive admittance at the center frequency.

Solution
The correct way to treat an apodized IDT is to divide it up into a number of channels in the x direction so that each channel is like an unapodized IDT. The admittances of all the channels are in parallel and can be added up. However, we can usually assume that the tap weights α_n alternate in sign and vary slowly in magnitude from one electrode to the next. In that case a good approximation for the capacitance is obtained if we treat it as an unapodized IDT with an effective number of electrodes N_{eff} given by

$$N_{\text{eff}} = \sum_n |\alpha_n| \tag{4.32}$$

The capacitance is then obtained from Example 4.17:

$$C_T = N_{\text{eff}} C_s W$$

In this case $N_{\text{eff}} = 2.5$, so that $C_T = 0.4$ pF and $2\pi f C_T = 1.26$ m. If the apodized IDT is built using double electrodes, the capacitance is 1.4 times higher (Example 4.18).

Actually, this concept of N_{eff} can also be used for the response function and the radiation conductance at the center frequency. Note from Example 4.14 that $\mu(f)$ at the center frequency for the apodized IDT is the same as that for an unapodized IDT with N_{eff} positive electrodes. This, of course, is true only at the center frequency. Since the passband shapes are totally different, we do not expect the equivalence to hold over a range of frequencies.

4.4. Model for Numerical Analysis

Let us briefly summarize what we have done to this point; this will serve to illustrate the limitations of analytical techniques and the need for a numerical model. In this chapter our purpose is to determine the admittance of the IDT as a function of frequency. The admittance has two parts: the ordinary capacitive part, which has nothing to do with surface waves, and the acoustic admittance, which arises from the interaction of the IDT and the surface waves generated by it (Fig. 4.3).

The real part of the acoustic admittance, called its *radiation conductance*, can be obtained from the response function of the IDT (Eq. 4.21). The response function, in general, is obtained from the summation in Eq. 4.17. For transducers with periodically spaced electrodes, the response function separates neatly into an element factor (Eq. 4.18, Fig. 4.9) that is known analytically and an array factor, which is the Fourier transform of the tap weights (Eq. 4.16 for unapodized IDTs, Eq. 4.19 for apodized IDTs). For apodized IDTs, the response function does not give the total radiation conductance; an additional component due to the nonuniform beam is present that depends on the beam profile (Eq. 4.28).

The imaginary part of the acoustic admittance, called its *radiation susceptance*, is obtained from the Hilbert transform of the radiation conductance (Eq. 4.29). It is zero at center frequency, changing from capacitive to inductive from lower to higher frequencies.

The preceding summary serves to illustrate the limitations on how far we can go with analytical methods, using the computer only to perform Fourier transforms. First, we are restricted to transducers with periodically spaced electrodes (this eliminates withdrawal-weighted IDTs). We have to neglect end effects; this is usually not very serious, especially if we add a couple of grounded electrodes at each end of the IDT.

Second, for apodized IDTs there is no simple way to get the total radiation conductance since the beam profile varies with frequency depending on the apodization. Third, we have no simple way to get the radiation susceptance. This does not matter for center frequency calculations (where it is zero), but it can affect the passband shape in strongly coupled IDTs, where it is comparable in magnitude to the capacitive susceptance.

Finally, there is one effect that we have consistently neglected: the reflections from the electrodes. As we pointed out before, these reflections cancel out at the center frequency for split-electrode IDTs. But they are significant for solid-electrode IDTs, which are often used in making high-frequency devices in order to save on photolithographic resolution. We like to have a general model for the IDT that can be readily implemented on a computer, which will give us the acoustic admittance as a function of frequency.

The transmission-line model we are going to develop now gives us just that. Consider the nth electrode of the IDT with all neighboring electrodes grounded (Fig. 4.24). We have an acoustic transmission line of impedance Z_0 coupled to the electrical busbars. The electrical line is described by its voltage V and current I, while the acoustic line is described by the wave amplitudes ϕ^+ and ϕ^- in the two directions. What we would like to do is

4 INTERDIGITAL TRANSDUCERS

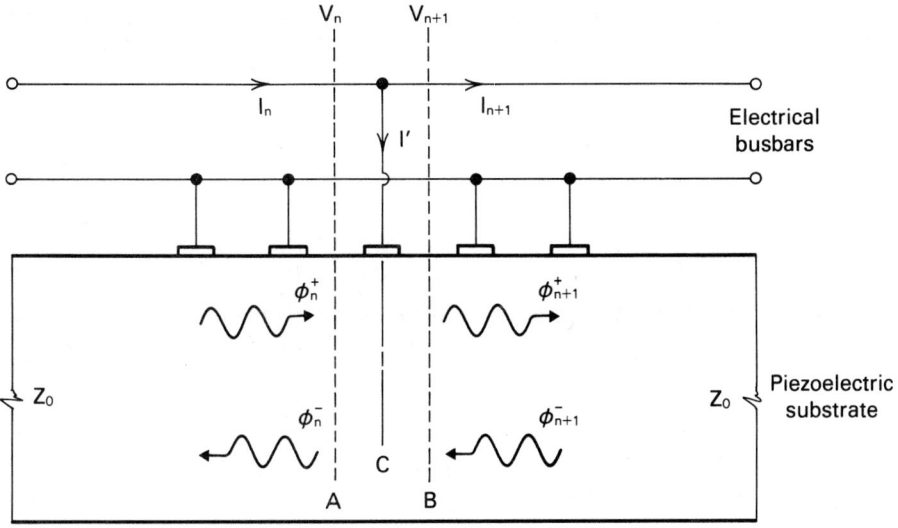

Figure 4.24 Modeling a single electrode in an IDT.

generate a transmission matrix $[T_n]$ for the nth electrode connecting the field quantities at its output to those at its input:

$$\begin{bmatrix} \phi^+ \\ \phi^- \\ V \\ I \end{bmatrix}_{n+1} = \begin{bmatrix} & & \\ & T_n & \\ & & \end{bmatrix} \begin{bmatrix} \phi^+ \\ \phi^- \\ V \\ I \end{bmatrix}_n \qquad (4.33)$$

Then we can cascade the transmission matrices of all the N electrodes to obtain a transmission matrix $[T]$ for the complete IDT:

$$[T] = [T_1][T_2] \cdots [T_{N-1}][T_N] \qquad (4.34)$$

This $[T]$ matrix relates the input quantities to the output quantities:

$$\begin{bmatrix} \phi^+ \\ \phi^- \\ V \\ I \end{bmatrix}_{N+1} = \begin{bmatrix} & & \\ & T & \\ & & \end{bmatrix} \begin{bmatrix} \phi^+ \\ \phi^- \\ V \\ I \end{bmatrix}_1 \qquad (4.35)$$

Once we have calculated this $[T]$ matrix numerically, we can obtain all the desired terminal quantities by applying the appropriate boundary conditions:

$$V_{N+1} = V_1 \tag{4.36a}$$

$$I_{N+1} = 0 \tag{4.36b}$$

These two equations are true for either a transmitter or a receiver. For a transmitter driven from a voltage source V with a source impedance Z_g,

$$\phi^-_{N+1} = \phi^+_1 = 0 \tag{4.37a}$$

$$V = V_1 + I_1 Z_g \tag{4.37b}$$

For a receiver excited by a forward-going wave and terminated in a load Z_L,

$$\phi^-_{N+1} = 0 \tag{4.38a}$$

$$V_1 + I_1 Z_L = 0 \tag{4.38b}$$

Using these equations, we can calculate all quantities of interest once we have the composite $[T]$ matrix. What we thus need is a recipe for writing down the transmission matrix for a single electrode. The computer can then perform the cascading operation (Eq. 4.34) and give us the composite transmission matrix. The transmission matrix for a single electrode can readily be written down in terms of the parameters μ and g_m that we have been using throughout this chapter. This is illustrated in the following examples.

Before we proceed let me point out a slight modification that we need in order to treat apodized IDTs. As we have seen before, apodized IDTs produce nonuniform beam profiles so that the wave amplitude $\phi^\pm$ varies along the width of the IDT. So apodized IDTs are usually divided up into channels along the width, so that the wave amplitude can be assumed constant in each channel (Fig. 4.25). Each channel then looks like an unapodized IDT with a different voltage sequence. The transmission-line model is used to find the acoustic admittance of each channel. Since all the channels are in parallel, the total admittance is the sum of the individual channel admittances.

4 INTERDIGITAL TRANSDUCERS

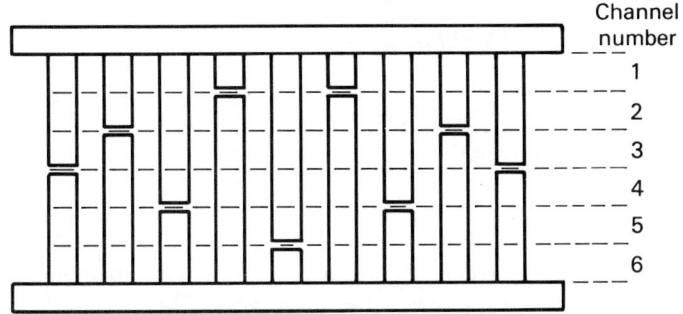

Figure 4.25 Apodized IDT divided into channels.

Example 4.20
Consider a single electrode in an IDT on Y-Z LiNbO$_3$ of width 350 μm with the electrodes spaced 17.5 μm periodically. Calculate its transmission matrix, neglecting reflections from electrodes.

Solution

$$v_0 = 3500 \text{ m/s} \quad \text{(Table 3.2)}$$

$$y_0 = 0.22 \text{ mmho}$$

$$p = 17.5 \text{ } \mu\text{m}$$

$$f_0 = \frac{v_0}{2p} = 100 \text{ MHz}$$

$$W = 350 \text{ } \mu\text{m}$$

$$\frac{W}{\lambda} = 10 \times \frac{f}{f_0}$$

For a single electrode the array factor is 0.5 for positive electrodes and -0.5 for negative electrodes (Eq. 4.16b). Hence its transmitter response function μ_0 is obtained from Eq. 4.16a:

$$\mu_0 = \pm 0.5 \text{ } \mu_s(f, \eta) \tag{4.39a}$$

where $\mu_s(f, \eta)$ is calculated from Eqs. 4.18a and b. Its receiver response function g_{m0} is obtained from μ_0 using Eq. 4.4:

$$g_{m0} = (4.44 \text{ mmho}) \mu_0 \frac{f}{f_0} \tag{4.39b}$$

Its radiation conductance G_0 is obtained using Eq. 4.21, noting that μ and g_m are always pure imaginary:

$$G_0 = -\mu_0 g_{m0} \tag{4.39c}$$

Now we can write down the transmission matrix defined in Eq. 4.35 (Fig. 4.24). We know from our earlier discussion that

$$\begin{bmatrix} \dfrac{\phi_n^-}{P} \\ \phi_{n+1}^+ P \\ I' \end{bmatrix} = \begin{bmatrix} 0 & 1 & \mu_0 \\ 1 & 0 & \mu_0 \\ -g_{m0} & -g_{m0} & G_0 \end{bmatrix} \begin{bmatrix} \phi_n^+ P \\ \phi_{n+1}^- P \\ V \end{bmatrix} \tag{4.40a}$$

where

$$P = e^{-j\pi f/2f_0} \tag{4.40b}$$

$$V = V_n = V_{n+1} \tag{4.40c}$$

$$I' = I_n - I_{n+1} \tag{4.40d}$$

Equation 4.40 follows from the definitions of μ_0 and g_{m0}. The phase factor P corresponding to acoustic propagation from A to C (or C to B) arises because μ and g_m are defined with the waves referenced to the electrode center, while in Fig. 4.24 the reference planes are shifted into the gap regions (plane A for ϕ_n's and plane B for ϕ_{n+1}'s). With some algebraic manipulation we can write Eqs. 4.40 in the form of a transmission matrix (Eq. 4.35):

$$\begin{bmatrix} \phi^+ \\ \phi^- \\ V \\ I \end{bmatrix}_{n+1} = \begin{bmatrix} P^2 & 0 & \mu_0 P & 0 \\ 0 & \dfrac{1}{P^2} & -\dfrac{\mu_0}{P} & 0 \\ 0 & 0 & 1 & 0 \\ g_{m0} P & \dfrac{g_{m0}}{P} & 0 & 1 \end{bmatrix} \begin{bmatrix} \phi^+ \\ \phi^- \\ V \\ I \end{bmatrix}_n \tag{4.41}$$

4 INTERDIGITAL TRANSDUCERS

μ_0 and g_{m0} are defined by Eqs. 4.39a and 4.39b and are both positive for positive electrodes, negative for negative electrodes.

Let us check the validity of Eq. 4.41. Suppose that we use the one-electrode IDT as a transmitter with an applied voltage V and zero generator impedance. Then using Eqs. 4.36 and 4.37, with a little algebra we get

$$\frac{\phi_1^-}{V} = \frac{-T_{23}}{T_{22}} = \mu_0 P$$

$$\frac{\phi_2^+}{V} = T_{13} = \mu_0 P$$

$$\frac{I_1}{V} = -T_{43} + \frac{T_{42} T_{23}}{T_{22}} = G_0 \qquad \text{(using Eq. 4.39)}$$

This is exactly what we would expect. The capacitive admittance $j2\pi f C_T$ should be calculated separately (Section 4.3.2) since it is not included in the numerical model.

Now, suppose that we use it as a receiver for a wave of amplitude ϕ_1^+ from the left. Using Eqs. 4.36 and 4.38 gives us

$$\frac{\phi_2^+}{\phi_1^+} = T_{11} = P^2$$

$$\frac{I_1}{\phi_1^+} = -T_{41} = -g_{m0} P$$

Again, this is what we expect. The negative sign in the induced current indicates that it goes into the generator.

In this example we assumed that the electrode was in a periodic environment so that Eqs. 4.47 can be used to calculate its response functions. Nonperiodic transducers can be analyzed by using the appropriate μ and g_m for each electrode in its environment.

Example 4.21
Consider the IDT in Example 4.20 but now with one positive and one negative electrode. Cascade transmission matrices of the form in Eq. 4.48 to calculate the acoustic admittance of the IDT.

Solution
The overall transmission matrix $[T]$ is obtained by cascading the two individual transmission matrices. For the negative electrode, μ and g_m are both negative.

$$T = \begin{bmatrix} P^2 & 0 & -\mu_0 P & 0 \\ 0 & \frac{1}{P^2} & \frac{\mu_0}{P} & 0 \\ 0 & 0 & 1 & 0 \\ -g_{m0}P & -\frac{g_{m0}}{P} & 0 & 1 \end{bmatrix} \begin{bmatrix} P^2 & 0 & \mu_0 P & 0 \\ 0 & \frac{1}{P^2} & \frac{\mu_0}{P} & 0 \\ 0 & 0 & 1 & 0 \\ g_{m0}P & \frac{g_{m0}}{P} & 0 & 1 \end{bmatrix}$$

$$= \begin{bmatrix} P^4 & 0 & \mu_0 P(P^2-1) & 0 \\ 0 & \frac{1}{P^4} & \frac{\mu_0(P^2-1)}{P^3} & 0 \\ 0 & 0 & 1 & 0 \\ g_{m0}P(1-P^2) & \frac{g_{m0}(P^2-1)}{P^3} & \frac{G_0(P^2-1)}{P^3} & 1 \end{bmatrix}$$

Consider the IDT used as a transmitter. Then, as in Example 4.20, we have

$$\mu^- = \frac{\phi_1^-}{V} = \mu_0 P(1 - P^2)$$

$$\mu^+ = \frac{\phi_3^+}{V} = \mu_0 P(P^2 - 1)$$

$$Y_a = \frac{I_1}{V} = 2G_0(1 - P^2)$$

Using Eq. 4.40b for P, the factor $(1 - P^2)$ is readily recognized as the array factor (Eq. 4.16). Note that the transmission-line model gives us both G_a and B_a:

$$G_a = 4G_0 \sin^2 \frac{\pi f}{2f_0}$$

4 INTERDIGITAL TRANSDUCERS

$$B_a = 2G_0 \sin \frac{\pi f}{f_0}$$

where G_0 is defined by Eq. 4.39c. As we would expect, B_a changes from capacitive to inductive at $f = f_0$. Also, it can be checked that G_a satisfies Eq. 4.21.

Example 4.22
If we compare Eq. 4.40 describing SAW generation by an IDT with Eq. 2.58 describing the in-line Mason model for bulk-wave generation, we can see two important differences. In the latter case, (1) we have $+\mu$ and $-\mu$ in row 1 and row 2 rather than $+\mu$ in both; and (2) μ is purely real rather than purely imaginary. Show that if instead of referencing the SAW amplitude to the electrode center (as we have done here), we reference it to the center of the gap between positive and negative electrodes (as in the in-line Mason model), both these differences are removed.

Solution
If we shift the reference from $z = 0$ (positive electrode) and $z = p$ (negative electrode) to $z = p/2$, then μ is multiplied by

$$e^{\pm j\pi f/2f_0} - e^{\mp j\pi f/2f_0} = \mp 2j \sin \frac{\pi f}{2f_0}$$

where the upper sign is for waves traveling in the $+z$ direction and the lower sign is for waves traveling in the $-z$ direction. This makes μ purely real and have opposite signs for waves in the $+z$ and $-z$ directions.

In our example we neglected reflections from the electrodes. These can be included in the transmission-line model by introducing a coupling between the forward and reverse waves. Equation 4.40a is then modified to

$$\begin{bmatrix} \dfrac{\phi_n^-}{P} \\ \dfrac{\phi_{n+1}^+}{P} \\ I' \end{bmatrix} = \begin{bmatrix} S_{11} & S_{12} & \mu_0 \\ S_{21} & S_{22} & \mu_0 \\ -g_{m0} & -g_{m0} & G_0 \end{bmatrix} \begin{bmatrix} \phi_n^+ P \\ \phi_{n+1}^- P \\ V \end{bmatrix} \qquad (4.42)$$

where S_{ij} are the scattering parameters of the shorted IDT ($V = 0$). In the absence of reflections, $S_{11} = S_{22} = 0$ and $S_{12} = S_{21} = 1$. But if each electrode has a reflection coefficient of r, then

$$S_{11} = S_{22} = r(1 + |r|^2)^{-1/2} \qquad (4.43\text{a})$$

$$S_{12} = S_{21} = (1 + |r|^2)^{-1/2} \qquad (4.43\text{b})$$

It should be noted that r is usually purely imaginary. With some algebra we can put Eq. 4.42 in the form of a transmission matrix (Eq. 4.35):

5

MULTISTRIP COUPLERS

A *multistrip coupler* (MSC) consists of a periodic array of isolated electrodes (Fig. 5.1); with proper design an MSC can be used to transfer a surface-wave beam completely from one track to another (Ref. 5.1). Such a track changer finds a variety of uses in surface-wave devices:

1. Track 2 can be designed to be narrower than track 1 so that the coupler can be used to compress a wide beam into a narrow one. This is useful in nonlinear devices that require high power density, such as convolvers.
2. The output SAW beam in track 2 is uniform, even if the incident wave in track 1 has a nonuniform profile. Thus the coupler can be used to cascade two apodized transducers; one is used to generate the incident wave in track 1, and the other receives the uniform SAW beam transferred to track 2 by the MSC.
3. The MSC transfers only the surface wave from track 1 to track 2; it does not transfer any other modes that may be present in track 1. Thus by placing the generator in track 1 and the receiver in track 2, spurious responses due to bulk mode generation can be eliminated.

The operation of the MSC can be understood in terms of the IDT theory discussed in Chapter 4. In Section 5.1 we analyze the effect of a single electrode (Fig. 5.2a) on an incident wave of amplitude ϕ_1 in track 1. The overall response of the coupler is then determined in Section 5.2 by cascading the response of all the electrodes. In this discussion we assume two tracks and identical electrodes, so that the cascading can be done analytically; however, nonperiodic multitrack couplers can be analyzed numerically in a straightforward manner (Ref. 5.2).

To design an MSC we need to specify the periodicity of the electrodes and the total number of electrodes. Usually, we choose $p = \lambda/4$ (with $\eta = 0.5$). This value is not critical as long as it is not close to $\lambda/2$; if it were close to $\lambda/2$, there would be strong reflections, which is not what we want. If the two tracks are of equal width, the number of electrodes N_T is given by

$$N_T = \frac{\pi}{\alpha} \tag{5.12}$$

where for $\eta = 0.5$,

$$\alpha = 0.64 K^2 \sin \frac{\pi f}{2f_0} \tag{5.4}$$

f_0 being the frequency at which the electrodes are spaced by $\lambda/2$ ($f_0 = v_0/2p$). The variation of α with η is shown graphically in Fig. 5.3; an approximate analytical expression is given in Eq. 5.7.

If the two tracks are of unequal widths then the number of electrodes N_T' is given by

$$N_T' = N_T \frac{W}{2\sqrt{W_1 W_2}} \tag{5.18}$$

where W_1 and W_2 are the track widths and W is the total coupler width. The period has to be made slightly smaller in the narrower track to achieve complete transfer.

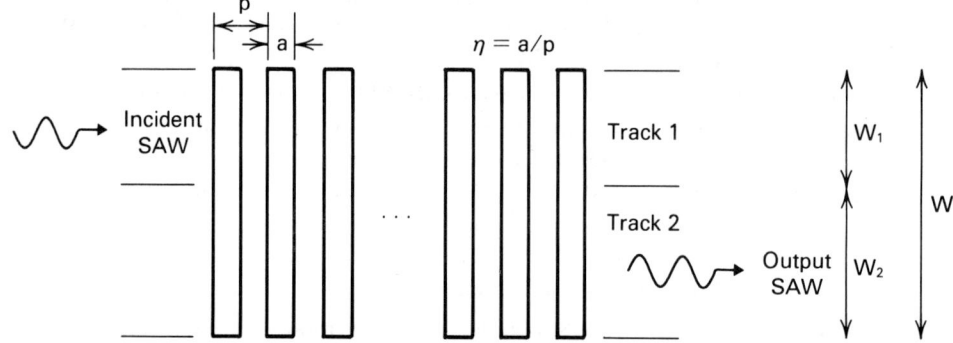

Figure 5.1 Multistrip coupler.

5 MULTISTRIP COUPLERS

$$\frac{\Delta}{p} = \frac{\alpha(W_1 - W)}{W} \frac{f_0}{\pi f} \tag{5.13}$$

where p and $p - \Delta$ are the periods of the wider and narrower track, respectively.

5.1. Coupling of Tracks by a Single Electrode

Consider an array of unconnected electrodes excited by an incident SAW in track 1. The potentials on the floating electrodes are all equal in magnitude, but their phase changes along the array to match the potential ϕ associated with the SAW. For example, if the strips were infinitesimally thin ($\eta \simeq 0$) and the incident wave were uniform over tracks 1 and 2 they would just float on the wave, so that the potential V_n on the nth electrode located at $z = z_n$ would be given by

$$V_n = \phi e^{-jkz_n}$$

$$= \phi e^{-jn\pi f/f_0}$$

where $\quad z_n = np$

$$f_0 = v_0/2p$$

For strips with a finite thickness we expect that the potential will be somewhat less. Calculated at $\eta = 0.5$ the potential is about 0.8ϕ (Ref. 5.6).

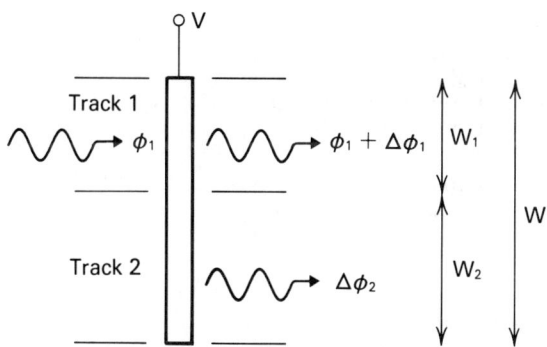

Figure 5.2 SAW beam incident in track 1 on a single electrode not shown.

$$V_n = 0.8 \frac{W_1}{W} \phi e^{-jn\pi f/f_0} \tag{5.1}$$

We have also added a factor W_1/W to account for the fact that the incident wave illuminates only track 1, which is a fraction W_1/W of the total width.

It is now straightforward to calculate the amplitudes $\Delta\phi_1$ and $\Delta\phi_2$ of the waves regenerated by each electrode in tracks 1 and 2 using the transducer theory discussed in Chapter 4.

$$\Delta\phi_1 = \Delta\phi_2 = 0.8 \frac{W_1}{W} \mu_s \phi \tag{5.2}$$

At $\eta = 0.5$, $\mu_s = 0.8 jK^2 \sin(\pi f/2f_0)$ (see Eq. 4.18a), so that we may write

$$\Delta\phi_1 = \Delta\phi_2 = j\alpha \frac{W_1}{W} \phi_1 \tag{5.3}$$

where

$$\alpha = 0.64 K^2 \sin \frac{\pi f}{2f_0} \tag{5.4}$$

Similarly, if we have an incident SAW of amplitude ϕ_2 in track 2, the regenerated wave will be given by

$$\Delta\phi_1 = \Delta\phi_2 = j\alpha \frac{W_2}{W} \phi_2 \tag{5.5}$$

Combining Eqs. 5.3 and 5.5 gives us

$$\Delta\phi_1 = j\alpha \left[\frac{W_1}{W} \phi_1 + \frac{W_2}{W} \phi_2 \right] \tag{5.6a}$$

$$\Delta\phi_2 = j\alpha \left[\frac{W_1}{W} \phi_1 + \frac{W_2}{W} \phi_2 \right] \tag{5.6b}$$

So far we have considered a metallization ratio η of 0.5. The coupling constant α is a maximum for this value of η. The overall variation in α over the range $0.25 < \eta < 0.75$ is shown in Fig. 5.3. At $\eta = 0.25$ or 0.75, α decreases by about 13% from its value at $\eta = 0.5$. We may write approximately

5 MULTISTRIP COUPLERS

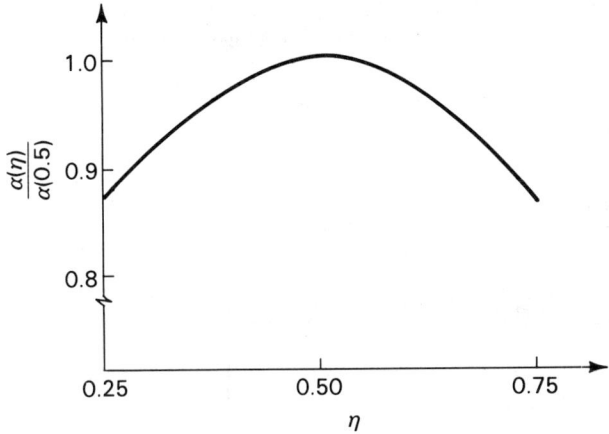

Figure 5.3 Variation of α with metallization ratio η.

$$\alpha(\eta) = \alpha(\eta = 0.5)\left[1 - 2(\eta - 0.5)^2\right] \qquad (5.7)$$

5.2. Overall Coupler Operation

As we have just seen, the surface waves in tracks 1 and 2 are coupled together by each electrode according to Eq. 5.6. Since K^2 is usually less than 5% or so, the coupling by a single electrode is rather small. So we can replace the $\Delta\phi$'s by differential $d\phi$ and write

$$\frac{d\phi_1}{dN} = j\alpha\left[\frac{W_1}{W}\phi_1 + \frac{W_2}{W}\phi_2\right] \qquad (5.8a)$$

$$\frac{d\phi_2}{dN} = j\alpha\left[\frac{W_1}{W}\phi_1 + \frac{W_2}{W}\phi_2\right] \qquad (5.8b)$$

where N is the number of strips. Of course, N can change only in units of 1, so that the differential seems odd; but since the change is small over a strip, this is not very serious. Strictly speaking, however, we should really solve a difference equation rather than a differential equation.

Equations 5.8 are a set of equations that occur in many physical situations, in fact, whenever two modes are coupled together, so that the behavior that we are about to see is typical of coupled modes. Consider what happens when a wave of amplitude ϕ_1 is incident on track 1. We would think that after a few strips the wave would divide equally between the two tracks. This is just what would happen if the j were absent from Eqs. 5.8. But the presence of the j makes a tremendous difference; the wave periodically shifts over from track 1 to track 2 and back. This can be seen by formally solving Eqs. 5.8a and 5.8b. Combining the two equations, we have

$$\frac{d^2\phi_1}{dN^2} = j\alpha \frac{d\phi_1}{dN} \tag{5.8c}$$

The solution to Eqs. 5.8 is given by

$$\phi_1(N) = \phi_1(0)\left[1 - \frac{W_1}{W}(1 - e^{j\alpha N})\right] \tag{5.9a}$$

$$\phi_2(N) = \phi_1(0)\left[\frac{W_1}{W}(1 - e^{j\alpha N})\right] \tag{5.9b}$$

Here we have assumed that at the input to the coupler (where $N = 0$) $\phi_2 = 0$. We can calculate the powers P_1 and P_2 in the two tracks from the amplitudes ϕ_1 and ϕ_2 using the relation $P \sim |\phi|^2$.

$$P_1(N) = P_1(0)\left[1 - \frac{2W_1W_2}{W^2}(1 - \cos \alpha N)\right] \tag{5.10a}$$

$$P_2(N) = P_1(0)\left[\frac{2W_1W_2}{W^2}(1 - \cos \alpha N)\right] \tag{5.10b}$$

Note that the power periodically transfers over from track 1 to track 2 and then back again. In designing a track changer, the trick is to truncate the coupler at $N = \pi/\alpha$ when the transfer to track 2 is maximum. At this point,

$$P_1(N = \frac{\pi}{\alpha}) = P_1(0)\left[\frac{W_1 - W_2}{W}\right]^2 \tag{5.11a}$$

$$P_2(N = \frac{\pi}{\alpha}) = P_1(0)\frac{4W_1W_2}{W^2} \tag{5.11b}$$

Note that if the two tracks are equal in width, the power is completely transferred to track 2 and the number of strips N_T needed for complete transfer is given by

5 MULTISTRIP COUPLERS

$$N_T = \frac{\pi}{\alpha} \tag{5.12}$$

where α is the coupling constant given by Eq. 5.7.

An important point to note is that the MSC is broadband. Any frequency dependence of the coupling comes from the element factor, α; the array factor is identically 1. This is because it is like a transducer in which the electrode voltages are induced by a wave traveling with the same velocity as the wave being generated; there is no interference between waves generated at different electrodes. The same factor also makes it unidirectional; the electrode voltages are phased as in a UDT. This is not true, however, if the frequency is equal to f_0. At this frequency the electrode voltages are phased by 180°, so that it also becomes bidirectional. In other words, it reflects the incident wave strongly because reflections from all the electrodes interfere constructively. So an MSC is never used around f_0, the frequency at which the strips are half a wavelength apart.

We have mentioned before that an important use of the multistrip coupler is to construct beam compressors where $W_2 < W_1$. But we have just seen that unless the tracks are of equal width, complete transfer of power is not possible. There is a way to get around this problem which we will now describe.

Physically, the reason for incomplete transfer is that the waves in the two tracks suffer different phase shifts as they traverse the coupler. So they get out of phase and coherent power transfer stops. For example, if we neglect the cross-coupling terms in Eq. 5.8, we have

$$\frac{d\phi_1}{dN} = j\alpha \frac{W_1}{W} \phi_1 \rightarrow \phi_1 \propto e^{j\alpha \frac{W_1}{W} N}$$

$$\frac{d\phi_2}{dN} = j\alpha \frac{W_2}{W} \phi_2 \rightarrow \phi_2 \propto e^{j\alpha \frac{W_2}{W} N}$$

The phase shifts in the two tracks are thus unequal unless $W_1 = W_2$. One way of explaining the difference in phase shifts is that both tracks are loaded by the same capacitance (that of the electrodes). But the wider track has a higher admittance (Y_0) and is less affected.

To get complete power transfer, we have to compensate this difference in phase shift. This is done by bending the electrodes so that they are closer together in the narrower track than in the wider one (Fig. 5.4). The phase delay due to propagation is e^{-jkp} in the wider track and $e^{-jk(p-\Delta)}$ in the narrower track. We have to choose Δ such that this exactly compensates the difference in phase shift discussed earlier:

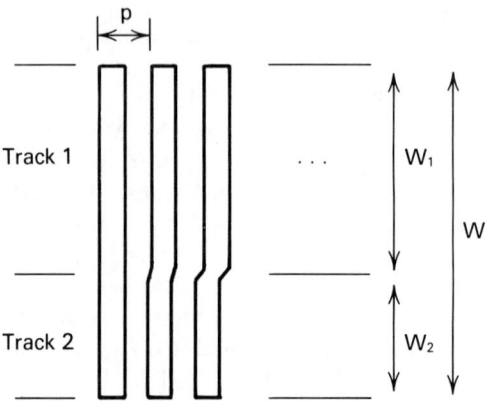

Figure 5.4 Multistrip beam compressor with bent electrodes.

$$\frac{\alpha W_1}{W} - kp = \frac{\alpha W_2}{W} - k(p - \Delta)$$

This gives

$$\frac{\Delta}{p} = \frac{\alpha(W_1 - W)}{W} \frac{f_0}{\pi f} \qquad (5.13)$$

where we have used the relation $kp = \pi f / f_0$.

At this point the reader might be puzzled by the sudden appearance of this new phase shift due to propagation delay; in Eqs. 5.8 we neglected this phase shift. This is apparent if we set $\alpha = 0$ (no piezoelectric coupling). Equations 5.8 then predict that $\phi_1, \phi_2 =$ constant, which is clearly wrong since for propagating waves we must have $\phi_1, \phi_2 \sim e^{-jkz}$. With $\alpha = 0$ we should get $\phi_1 \sim e^{-jkp_1 N}$ and $\phi_2 \sim e^{-jkp_2 N}$, where p_1 and p_2 are the spacing between successive strips in tracks 1 and 2, respectively. Equations 5.8 should actually read

$$\frac{d\phi_1}{dN} = j\left[\frac{\alpha W_1}{W} - kp_1\right]\phi_1 + \frac{j\alpha W_2}{W}\phi_2$$

$$\frac{d\phi_2}{dN} = j\frac{\alpha W_1}{W}\phi_1 + j\left[\frac{\alpha W_2}{W} - kp_2\right]\phi_2$$

The reason we did not bother about these extra terms is that as long as $p_1 = p_2$, the extra phase shifts are the same for both tracks 1 and 2.

5 MULTISTRIP COUPLERS

Coupling between two waves is not affected at all by equal phase shifts; it is the relative phase shift that matters. Now if we choose $p_1 = p$ and $p_2 = p - \Delta$ such that Eq. 5.13 satisfied, we can write

$$\frac{d\phi_1}{dN} = j\left[C\phi_1 + \frac{\alpha W_2}{W}\phi_2\right] \tag{5.14a}$$

$$\frac{d\phi_2}{dN} = j\left[\frac{\alpha W_1}{W}\phi_1 + C\phi_2\right] \tag{5.14b}$$

where $C = (\alpha W_1/W) - kp = (\alpha W_2/W) - k(p - \Delta)$. As long as we have the same C in both equations, its precise value is unimportant, since it causes the same phase shift for both waves; this can be seen easily by transforming to a new set of variables $\phi'_{1,2} = \phi_{1,2} \exp(-jCN)$. We will then find that the new equations in terms of ϕ' are the same as Eq. 5.14 with $C = 0$. For simplicity, let us just set $C = 0$; Eqs. 5.14a and 5.14b are then readily combined to yield

$$\frac{d^2\phi_1}{dN^2} = -\alpha^2 \frac{W_1 W_2}{W^2}\phi_1 \tag{5.14c}$$

The solution to Eqs. 5.14 is given by

$$\phi_1(N) = \phi_1(0) \cos \alpha' N \tag{5.15a}$$

$$\phi_2(N) = j\frac{\sqrt{W_2}}{W_2}\phi_1(0) \sin \alpha' N \tag{5.15b}$$

where

$$\alpha' = \alpha \frac{\sqrt{W_1 W_2}}{W} \tag{5.16}$$

The powers P_1 P_2 in the two tracks are written as

$$P_1(N) = P_1(0) \cos^2 \alpha' N \tag{5.17a}$$

$$P_2(N) = P_1(0) \sin^2 \alpha' N \tag{5.17b}$$

showing complete power transfer when the number of electrodes N'_T is equal to $\pi/2\alpha'$. Comparing with Eq. 5.12, we see that

$$N_T' = N_T \frac{W}{2\sqrt{W_1 W_2}} \tag{5.18}$$

where we have used Eq. 5.16 for α'. The number of electrodes needed for complete transfer is thus larger than that needed in the case of equal-width tracks by a factor of $W/2\sqrt{W_1 W_2}$.

Example 5.1
Design a multistrip coupler for transferring surface waves between tracks of equal width, to operate around a center frequency of 100 MHz on Y-Z lithium niobate.

Solution
We need to determine the number of electrodes N and the spacing, p between them. The spacing p is not critical as long as it is not close to $\lambda/2$. A good number to use is

$$p = \frac{\lambda}{4} \quad \text{so that the center frequency } f_c = \frac{f_0}{2}$$

Also,

$$N = \frac{\pi}{\alpha} \tag{5.12}$$

$$\alpha = 0.46 \, K^2 \quad \text{at } f_c = \frac{f_0}{2} \tag{5.7}$$

For Y-Z lithium niobate,

$$v_0 = 3500 \; m/s$$

$$K^2 = 4.6\%$$

$$\lambda = 35 \; \mu m \quad \text{at } f_c = 100 \; \text{MHz}$$

$$\alpha = 0.019$$

$$p = 8.75 \; \mu m$$

$$N = 165$$

5 MULTISTRIP COUPLERS

Note that the multistrip coupler is fairly broadband. The frequency response of the coupler comes from the frequency dependence of α. From Eq. 5.10,

$$P_2(N) = P_1(0) \sin^2 \frac{\alpha N}{2}$$

where

$$\alpha = 0.64\, K^2 \sin \frac{\pi f}{2 f_0}$$

Example 5.2
Design a multistrip beam compressor on Y-Z lithium niobate to operate around 100 MHz with a compression ratio of 10:1.

Solution
As in Example 5.1,

$$p = 8.75\ \mu\text{m}$$

We need to determine the number of electrodes N and the degree of bending Δ required.

$$N = 238 \tag{5.18}$$

$$\frac{\Delta}{p} = 0.012 \tag{5.13}$$

$$\Delta = 0.1\ \mu\text{m}$$

In this section we have assumed two tracks and periodic electrodes. More general cases of multitrack nonperiodic couplers can be numerically analyzed in a straightforward manner; we will briefly describe the approach in this section. Let ϕ_i^N denote the SAW amplitude in the Ith track at the Nth electrode. We assume that the period p is not close to $\lambda/2$, so that reflected waves can be neglected; we need to worry about waves in only one direction.

What we need for numerical analysis is a transmission matrix $T_{(N)}$ that connects $\phi_j^{(N)}$ to $\phi_i^{(N+1)}$.

$$\phi_i^{(N+1)} = T_{ij}^{(N)}\phi_j^{(N)} \tag{5.19}$$

We can then cascade all the transmission matrices to obtain an overall transmission matrix for the coupler. Each electrode can be different; there is no need for the $T^{(N)}$'s to be identical.

The expression for the transmission matrix for a single electrode is obtained readily by generalizing Eq. 5.6:

$$T_{ij}^{(N)} = \left[\delta_{ij} + j\,\alpha\frac{W_j}{W}\,e^{-j\theta_i}\right] \tag{5.20}$$

where θ_i is the phase shift due to the propagation delay in the ith track from one electrode to the next (center to center); θ_i is different for different tracks if the periods are different.

6
REFLECTORS

A reflector is basically a shorted interdigital transducer, that is, a periodic array of electrodes all electrically connected together (Fig. 6.1a). It reflects an incident surface wave completely over a narrow band of frequencies around f_0, the frequency at which the electrode spacing is half a wavelength. In the case of transducers and multistrip couplers, it is precisely this reflection that we try to avoid by operating at frequencies away from f_0 (commonly at $f_0/2$). Actually, a reflector can be made of unconnected electrodes, too, like a multistrip coupler. In this case there will also be an induced voltage on the electrodes so that an additional regeneration reflection (see Section 9.3.1) has to be considered. However, unconnected electrodes seem to show more spurious effects, so that practical reflectors always employ shorted electrodes. Reflectors are also made by etching grooves on the surface. These can alternatively be viewed as reflecting strips made of the substrate material.

Surface-wave reflectors are distributed reflectors; each strip has a small reflectivity r. Around the center frequency the strips are $\lambda/2$ apart and the reflections from different strips add up in phase. The array reflection coefficient R at center frequency is given by

$$R \simeq \tan Nr \tag{6.4c}$$

or $|R| \simeq \tanh(N|r|)$ (since r is pure imaginary), where N is the number of strips in the array. Note that R is very nearly equal to 1 if

$N|r| \gg 1$, indicating almost 100% reflectivity. The fractional bandwidth of the reflector, B/f_o, is given by Eq. 6.5. Here B is defined as the separation between the null frequencies.

$$\frac{B}{f_o} = \frac{2|r|}{\pi} \sqrt{1 + (\frac{\pi}{N|r|})^2} \qquad (6.5)$$

$$\simeq \frac{2|r|}{\pi} \qquad N|r| \gg \pi$$

There is one other important parameter that we like to know about a reflective array — the effective center of reflection L_p. The surface waves penetrate a certain distance into the array so that the reflector appears to be effectively located at a distance L_p from the leading edge (Figs. 6.4 to 6.6), which can be determined from the slope of the phase of R at center frequency:

$$L_p \simeq 1/4 \, |r| \qquad (6.7)$$

L_p is very important in determining the effective cavity length when two reflectors are used to form a resonator (Chapter 10).

In Section 6.1 we discuss the reflectivity, reflection bandwidth, and effective center of reflection of a reflector array assuming that each strip has a reflectivity of r. Section 6.2 describes how r can be determined for different kinds of reflectors, starting from the material parameters of the electrode and the substrate. We also discuss the change in SAW velocity produced by the strips; this gives rise to a slight shift in the center frequency of the reflector, which can be important in resonator design. Table 6.1 gives the material parameters for some common substrates and electrode materials, which can be used in Eqs. 6.9 and 6.10 to determine r and v. It should be mentioned that the results of Section 6.2 which allow the reflectivity per electrode r to be calculated from first principles, are fairly recent and have not yet been verified conclusively by experiments. However, they agree with the available experimental data and we felt that the inclusion of these results will be useful despite the lack of complete verification. The detailed theoretical derivation involves acoustic field theory beyond the scope of this book. In Section 6.2 we have presented the end results in a form that can be used easily by a device designer. In Section 6.3 we justify some of the results used in Section 6.2 with simple heuristic arguments based on an analogy with transmission lines. This section can be skipped on first reading.

6 REFLECTORS

TABLE 6.1: ρ, α_x, and α_z for Common Electrodes

Electrode Material	ρ (kg/m^3)	α_x (N/m^2)	α_z (N/m^2)
Aluminum	2,695	2.5×10^{10}	7.8×10^{10}
Chromium	7,200	10^{11}	28.2×10^{10}
Titanium	4,500	4.4×10^{10}	12.9×10^{10}
Gold	19,300	2.85×10^{10}	9.8×10^{10}
Silver	19,490	2.7×10^{10}	8.7×10^{10}
Y-Z LiNbO$_3$	4,700	—	20.9×10^{10}
128°-rotated Y-cut LiNbO$_3$ (x prop.)	4,700	7.7×10^{10}	18.6×10^{10}
ST-X quartz	2,651	6.7×10^{10}	8.7×10^{10}
77.5°-rotated Y-cut LiTaO$_3$ (perp. to x prop.)	7,450	—	22.4×10^{10}
100-cut 011 propagation GaAs	5,307	—	28.2×10^{10}

6.1. Operation of a Reflector Array

6.1.1. Transmission Matrix Formulation

The operation of a reflector array is analyzed in a rather straightforward manner. It can be modeled as a transmission line with a periodic impedance mismatch (Fig. 6.1b). There is also a susceptance jB at each junction, which does not affect the reflection at center frequency if $\eta = 0.5$. Let us neglect this for the moment; the origin and effect of this susceptance are discussed in Section 6.2 after Eq. 6.9b. The reflection r at every electrode, referenced to the center of the electrode, is given by (see Section 6.3.1 for more detail)

$$r = j \frac{\Delta Z}{Z} \sin \frac{\eta \pi f}{f_o} \quad (6.1a)$$

The center frequency f_o is determined by the average velocity of the wave, v, which is slightly different from the free surface velocity v_o. If v_e is the effective velocity under an electrode,

$$\frac{v - v_o}{v} = \frac{v_e - v_o}{v_e} \eta \quad (6.1b)$$

In Section 6.2 we discuss the factors that affect r and v; but let us first see how the reflector operation is analyzed if we do know r and v.

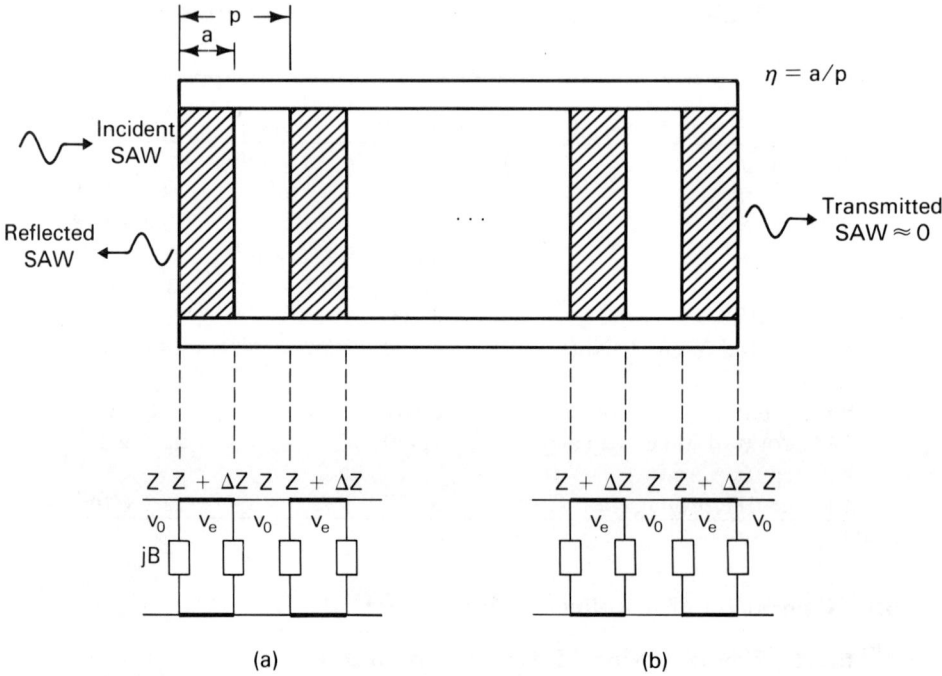

Figure 6.1 Surface wave reflector: (a) electrode configuration; (b) equivalent transmission-line model.

The forward- and backward-wave amplitudes ϕ^+ and ϕ^- at the nth and $(n+1)$st electrode (Fig. 6.2) are written as

$$\phi^+_{n+1} = e^{-jkp}\phi^+_n + re^{-j2kp}\phi^-_{n+1} \tag{6.2a}$$

$$\phi^-_n = r\phi^+_n + e^{-jkp}\phi^-_{n+1} \tag{6.2b}$$

where $k = 2\pi/\lambda$.

Equations 6.2 can be rearranged into a transmission matrix of the form discussed earlier in connection with the numerical modeling of IDTs (Eq. 4.41) (neglecting terms of order r^2).

$$\begin{Bmatrix}\phi^+ \\ \phi^-\end{Bmatrix}_{n+1} = \begin{bmatrix} P^2 & rP^2 \\ \dfrac{-r}{P^2} & \dfrac{1}{P^2}\end{bmatrix}\begin{Bmatrix}\phi^+ \\ \phi^-\end{Bmatrix}_n \tag{6.3}$$

where $P^2 = e^{-jkp} = e^{-j\pi f/f_o}$ and $f_o = v/2p$.

6 REFLECTORS

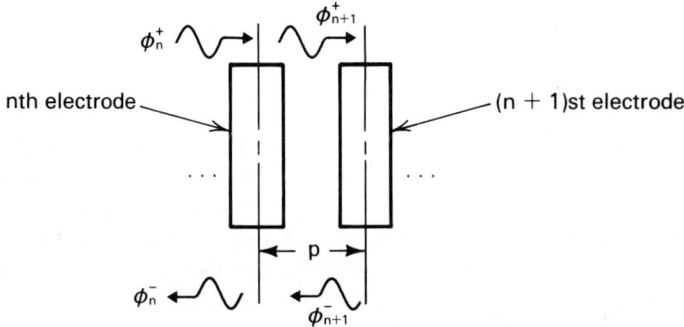

Figure 6.2 Forward and backward waves at the nth and $(n+1)$st electrodes.

Equation 6.3 gives the transmission matrix $[t]$ of a single electrode; successive transmission matrices can be cascaded numerically to give the overall transmission matrix for the array. The reflector characteristics are then readily reduced. For example, if $[T]$ represents the transmission matrix of an array of N electrodes, we have

$$\begin{bmatrix} \phi^+ \\ \phi^- \end{bmatrix}_{N+1} = \begin{bmatrix} T_{11} & T_{12} \\ T_{21} & T_{22} \end{bmatrix} \begin{Bmatrix} \phi^+ \\ \phi^- \end{Bmatrix}_1$$

Assuming an incident wave only from the left (that is, $\phi^-_{N+1} = 0$), we can calculate the reflected and transmitted powers.

$$P_R(\text{dB}) = 20 \log \frac{|\phi^-_1|}{|\phi^+_1|} = 20 \log \left| \frac{T_{21}}{T_{22}} \right| \tag{6.4a}$$

$$P_T(\text{dB}) = 20 \log \left| \frac{\phi^+_{N+1}}{\phi^+_1} \right| = 20 \log \left| \frac{T_{11}T_{22} - T_{12}T_{21}}{T_{22}} \right| \tag{6.4b}$$

It can be shown (see Example 6.1) that at the center frequency ($P^2 = -1$) the array reflectivity R is given by

$$R = \frac{-T_{21}}{T_{22}} \simeq \tan Nr \tag{6.4c}$$

or

$$|R| \simeq \tanh N|r|$$

where N is the total number of strips.

Example 6.1
Derive Eq. 6.4c.

Solution
As we have just discussed, we have to raise the transmission matrix $[t]$ for a single electrode to the power N to get the transmission matrix $[T]$ for the entire array; in general, this is done numerically on a computer. At the center frequency with $P^2 = -1$, we can calculate $[t]^N$ analytically using a little trick. Letting $P^2 = -1$ in Eq. 6.3 gives us

$$\begin{Bmatrix} \phi^+ \\ \phi^- \end{Bmatrix}_{n+1} = \begin{bmatrix} -1 & -r \\ +r & -1 \end{bmatrix} \begin{Bmatrix} \phi^+ \\ \phi^- \end{Bmatrix}_n$$

Let us define a new set of variables ϕ', ϕ'' as follows:

$$\phi' = \phi^+ + j\phi^-$$

$$\phi'' = \phi^+ - j\phi^-$$

It is easy to show that

$$\begin{Bmatrix} \phi' \\ \phi'' \end{Bmatrix}_{n+1} = \begin{bmatrix} -1 + jr & 0 \\ 0 & -1 - jr \end{bmatrix} \begin{Bmatrix} \phi' \\ \phi'' \end{Bmatrix}_n$$

The advantage of going to this new set of variables is now apparent; the transmission matrix is now diagonal. (This process for transforming variables is known as *diagonalizing* a matrix.) Diagonal matrices are readily raised to any power:

$$\begin{Bmatrix} \phi' \\ \phi'' \end{Bmatrix}_{N+1} = \begin{bmatrix} (-1 + jr)^N & 0 \\ 0 & (-1 - jr)^N \end{bmatrix} \begin{Bmatrix} \phi' \\ \phi'' \end{Bmatrix}_1$$

$$\simeq \begin{bmatrix} -e^{-jNr} & 0 \\ 0 & -e^{jNr} \end{bmatrix} \begin{Bmatrix} \phi' \\ \phi'' \end{Bmatrix}_1$$

We have used the approximation $e^x \simeq 1 + x$ for small x. Now we can go back to our usual variables ϕ^+ and ϕ^-:

$$\begin{Bmatrix} \phi^+ \\ \phi^- \end{Bmatrix}_{N+1} \simeq \begin{bmatrix} -\cos Nr & -\sin Nr \\ \sin Nr & -\cos Nr \end{bmatrix} \begin{Bmatrix} \phi^+ \\ \phi^- \end{Bmatrix}_1$$

6 REFLECTORS

Hence

$$R = -\frac{T_{21}}{T_{22}} \simeq \tan Nr$$

Since r is pure imaginary and $\tanh jx = j \tan x$, we have

$$|R| \simeq \tanh N|r|$$

6.1.2. Reflector Bandwidth

Figure 6.3 shows the reflection magnitude for a reflector array with 100, 200, and 400 strips each having a reflectivity $|r| = 0.01$. These have been determined numerically by cascading transmission matrices. As we can see, the reflectivity is 100% over a band of frequencies when $N = 400$. Physically, we can estimate the bandwidth of reflection with a simple argument. The frequency response of the reflectivity arises from the distributed nature of the reflector; reflections from successive electrodes must add in phase to generate a strong reflection. We would thus expect that the longer the reflector, the smaller the bandwidth; however, it is not

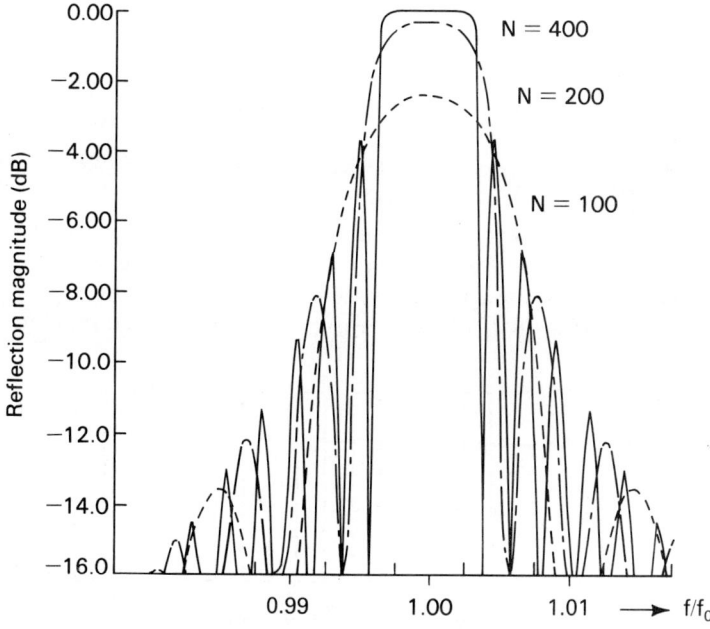

Figure 6.3 Reflection magnitude versus frequency ($r = 0.01$).

the physical length of the reflector that matters (provided that it is long enough to give approximately 100% reflectivity at the center frequency) since the wave is almost completely reflected by the first n strips, where n is $\sim 1/|r|$, $|r|$ being the reflectivity of each strip. So we would expect the fractional bandwidth

$$\frac{B}{f_0} \sim \frac{1}{N}|r|$$

An exact analysis gives

$$\frac{B}{f_0} = \frac{2|r|}{\pi}\sqrt{1 + \left(\frac{\pi}{N|r|}\right)^2} \tag{6.5}$$

where N is the number of strips. Note that the bandwidth is independent of N if $N|r| \gg \pi$; but if $N|r| \ll \pi$, it is independent of r.

$$\frac{B}{f_0} = \frac{2|r|}{\pi} \qquad N|r| \gg \pi$$

$$= \frac{2}{N} \qquad N|r| \ll \pi$$

Here B is defined as the *separation between the null frequencies* on either side of the center frequency.

6.1.3. Effective Center of Reflection

The wave decays exponentially into the reflector array, the decay rate being proportional to the reflectivity of each electrode. Conceptually, the reflector can be viewed as being located at an effective distance L_p from the leading edge (Fig. 6.4). This is very useful in the design of resonators

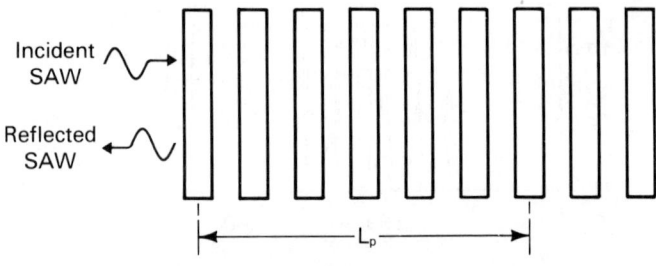

Figure 6.4 Effective center of reflection.

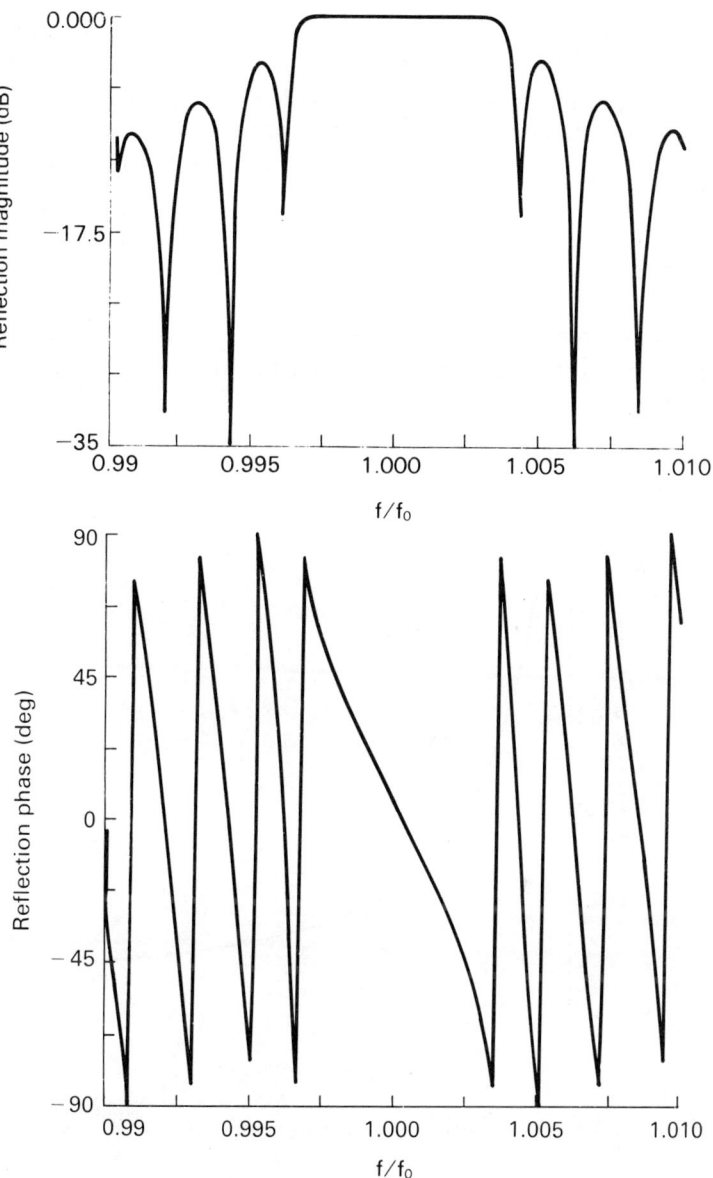

Figure 6.5   Reflection magnitude and phase versus frequency for $N = 400$, $r = 0.01$.

in which a cavity is formed by two reflectors and it is important to know the effective cavity length which determines the frequency spacing between cavity modes.

We expect L_p to be inversely proportional to r. The right way to determine L_p is by determining the phase of the reflected wave and looking at the phase slope at the center frequency. Figure 6.5 shows the reflection magnitude and phase for a reflector with $N = 400$, $r = 0.01$, numerically determined by cascading transmission matrices. The effective center of reflection can be determined from the phase slope using

$$L_p(\text{wavelengths}) = \frac{f_o}{4\pi} \left| \frac{d\theta}{df} \right| \qquad (6.6)$$

where θ is measured in radians. Figure 6.6 shows L_p determined for different values of r. The curve is described quite accurately by

$$L_p(\text{wavelengths}) = \frac{1}{4} |r| \qquad (6.7)$$

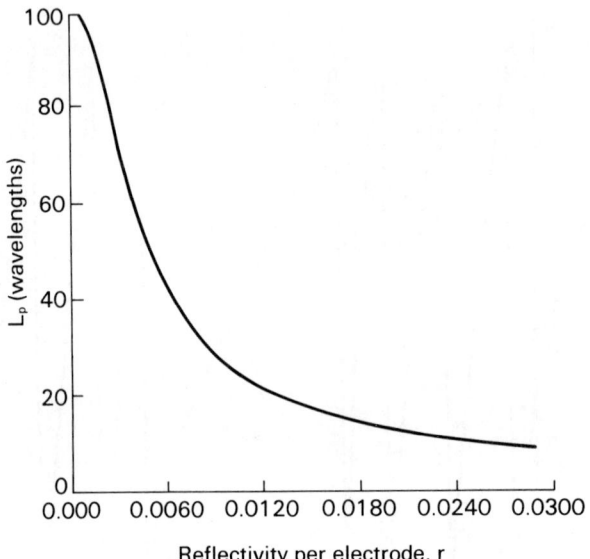

Figure 6.6 Effective center of reflection versus reflectivity per electrode.

Example 6.2
Consider a reflector with $f_o = 100$ MHz, $r = 0.02j$, and $N = 100$. Calculate (a) the bandwidth; (b) the effective position of the reflector; (c) the array reflection; (d) the transmission matrix of each electrode

Solution

$$(a) \; B \quad = 0.024 f_o = 2.4 \text{ MHz} \qquad (6.5)$$

$$(b) \; L_p \quad = 12.5 \; \lambda \qquad (6.7)$$

6 REFLECTORS

(c) $|R| = \tanh N|r| = 96.4\%$

(d) $P^2 = e^{-j\pi f/100 \text{ MHz}}$

So the transmission matrix of each electrode is given by (Eq. 6.3)

$$[t] = \begin{bmatrix} e^{-j\pi f/100 \text{ MHz}} & 0.02 je^{-j\pi f/100 \text{ MHz}} \\ -0.02 je^{j\pi f/100 \text{ MHz}} & e^{j\pi f/100 \text{ MHz}} \end{bmatrix}$$

For the overall array, $[T] = [t]^{100}$.

6.2 Reflection and Velocity Change Due to a Single Electrode

We have seen in Section 6.1 that to analyze a reflector array, we need two parameters: the reflectivity r and the velocity v. The velocity v differs by a small percentage from the free surface velocity v_o; but it is important to know the precise velocity since it determines the center frequency of operation. Reflectors are used chiefly to make resonators which are very narrowband, and even small uncertainties in their center frequency is serious.

In this section we will discuss how the reflectivity r and the fractional velocity change $(v - v_o)/v_o$ can be determined from a knowledge of the substrate material and the electrode material. A detailed analysis of these effects requires field theory at a level beyond the scope of this book, so we will merely describe the results with plausibility arguments and illustrative examples.

Before we proceed, let use look at the convention we will follow regarding the reflection coefficient r. We know that in transmission lines the conventional reflection coefficient is really the voltage reflection coefficient; the current reflection coefficient has the opposite sign. A surface wave is far more complicated, with a number of quantiites, such as particle velocities, stresses, and so on, combining to give the total flow of power. Some of these quantities are voltage-like (have the same sign for forward and reverse waves) and some are current-like (have opposite signs for forward and reverse waves). It is important to mention which quantity is used to determine the reflection coefficient. Since we have been using the surface potential ϕ as the amplitude, the reflection coefficient is referenced to ϕ. As we have mentioned before, ϕ is usually the field quantity of most interest since interdigital transducers detect it.

It is also important to specify the reference plane for the reflection coefficient. We will use the center of the electrode as our reference plane (Fig. 6.7). For grooved reflectors the reference will be the center of the step rather than the groove. The reflection coefficient at any other reference plane $z = z_o$ can be obtained easily.

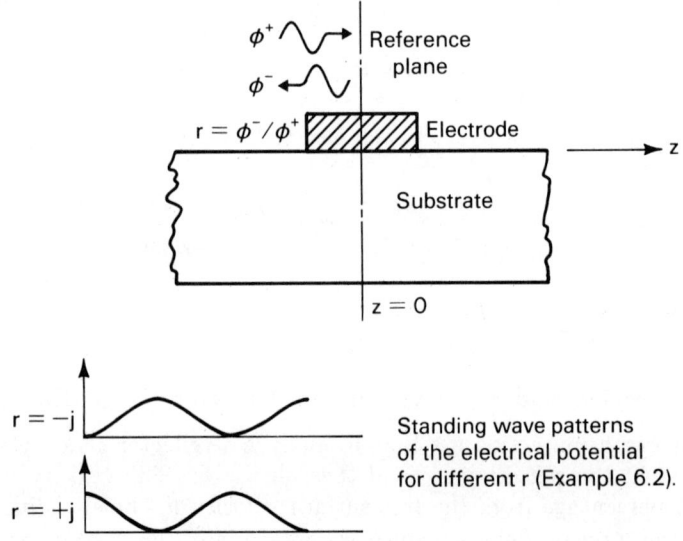

Figure 6.7 Reference plane for the reflection coefficient.

$$r(z = z_0) = r(z = 0)e^{j4\pi z_0/\lambda} \qquad (6.8)$$

Example 6.3
Consider a reflector array with each electrode having a reflectivity $r = -0.02j$ at the center frequency f_0 (metallization ratio = 0.5). (a) Calculate the fractional impedance mismatch $\Delta Z/Z$ in the transmission-line model. (b) What is the reflectivity referenced to the leading edge of the electrode? (c) A standing wave is formed due to the reflection of an incident wave from the array. How far from the first electrode is the nearest maximum of the surface potential? (This is very important in designing resonators, since the transducers have to be located at the potential maxima.)

Solution
(a) For a transmission line, $r = j\Delta Z/Z$ if $\eta = 0.5$ (Eq. 6.1a). Thus

$$\frac{\Delta Z}{Z} = -0.02$$

(b) The leading edge of the electrode is located at $z = -\lambda/8$ if the center is at $z = 0$. Hence from Eq. 6.8,

6 REFLECTORS

$$r(z = -\frac{\lambda}{8}) = r(z = 0)e^{-j\pi/2}$$

$$= -0.02$$

(c) Since the reflection coefficient is real and negative at the leading edge of the electrode, this must be the location of the potential minimum. The potential maximum will be a quarter-wavelength from the leading edge, where the reflection coefficient is real and positive (see Fig. 6.7).

We will now describe how the reflectivity and the fractional velocity change can be determined from the substrate and electrode material parameters. Both these quantities can be divided into two components: the piezoelectric and the mechanical. The former depends on the substrate coupling coefficient K^2, and the latter depends on the thickness h of the electrode in wavelengths. For grooved reflectors the piezoelectric part is absent. The reflectivity and velocity shift can be written in terms of four parameters, P_z, F_z, P_v, and F_v, which we will discuss shortly.

$$\frac{v - v_o}{v_o} = \left[P_v(\eta) \frac{K^2}{2} + F_v \frac{h}{\lambda} \right] \eta \qquad (6.9a)$$

$$r = j \left[P_z(\eta) \frac{K^2}{2} + F_z \frac{h}{\lambda} \right] \sin \frac{\eta \pi f}{f_o} \qquad (6.9b)$$

Note that the piezoelectric coefficients P_z and P_v depend on the metallization ratio; this is because the piezoelectric effect involves long-range fields that depend on neighboring electrode configurations. The first-order mechanical effect is represented by F_v and F_z. The velocity shift is also affected by what is known as the stored energy effect, which is neighbor dependent; there is no simple analytical expression for it. We will not discuss it further, although it can be of significance in determining $(v - v_o)/v_o$, especially when P_v and F_v are negligible, as in the case of grooved arrays. This effect is often modeled by the susceptance jB, shown in Fig. 6.1. Luckily, it does not affect the reflectivity at f_o if $\eta = 0.5$ (it gives rise to a reflection at $2f_o$, but this is not important in practical devices). It should be noted that the linear dependence of r with h/λ is not valid when h/λ is large (>2%). However, practical reflectors use small h/λ to reduce losses due to mode conversion.

Figure 6.8 shows $P_V(\eta)$ and $P_Z(\eta)$ plotted in the range $0.25 < \eta < 0.75$. These can be used to calculate the piezoelectric part of the reflection and the velocity change. The mechanical components F_z and F_v are written as follows:

$$F_v = \frac{K^2}{2C_s}\left[|c_x|^2(\alpha_x - \rho v_o^2) - |c_y|^2 \rho v_o^2 + |c_z|^2(\alpha_z - \rho v_o^2)\right] \quad (6.10\text{a})$$

$$F_z = \frac{-\pi K^2}{C_s}\left[c_x^2(\alpha_x + \rho v_o^2) + c_y^2 \rho v_o^2 + c_z^2(\alpha_z + \rho v_o^2)\right] \quad (6.10\text{b})$$

where K^2, C_s, and $c_{x,y,z}$ are properties of the substrate given in Tables 3.1 and 3.2, while ρ, α_x, and α_z describe properties of the electrode material. The former is just the mass-density, while α_z and α_x have the dimensions of stiffness and depend on various components of the stiffness tensor $[c]$. If the material is isotropic with Lamé constants λ and μ,

$$\alpha_x = \mu \quad (6.11\text{a})$$

$$\alpha_z = \frac{4\mu(\lambda + \mu)}{\lambda + 2\mu} \quad (6.11\text{b})$$

For an anisotropic material with a compliance tensor $[s]$ (inverse of $[c]$),

$$\alpha_x = \frac{S_3^3}{D} \quad (6.12\text{a})$$

$$\alpha_Z = \frac{S_5^5}{D} \quad (6.12\text{b})$$

where $S_I^J = (s_{11}s_{IJ} - s_{1J}s_{I1})/s_{11}$

$$D = S_5^5\, S_3^3 - S_3^5 S_5^3$$

Table 6.1 lists ρ, α_x, and α_z for some common electrode materials and also for some common substrate materials (grooved reflectors may be viewed as electrodes made of the substrate material). In Section 6.3 we present a heuristic derivation of Eq. 6.10 and discuss the similarities and differences of SAW reflections with reflections in a transmission line.

Example 6.4
Calculate the reflectivity per electrode for (a) aluminum reflectors on ST-X quartz; (b) grooved array on ST-X quartz; (c) aluminum reflectors on Y-Z lithium niobate; (d) gold reflectors on ST-X quartz.

6 REFLECTORS

Solution

(a) For ST-X quartz (Tables 3.1 and 3.2),

$$-\frac{\pi K^2}{C_s} c_x^2 = 2.5 \times 10^{-13} \text{ m}^2/\text{N}$$

$$-\frac{\pi K^2}{C_s} c_y^2 = 2.8 \times 10^{-11} \text{ m}^2/\text{N}$$

$$-\frac{\pi K^2}{C_s} c_z^2 = -1.2 \times 10^{-11} \text{ m}^2/\text{N}$$

For aluminum reflectors (Table 6.1),

$$\alpha_x = 2.5 \times 10^{10} \text{N}/\text{m}^2$$

$$\alpha_z = 7.8 \times 10^{10} \text{ N}/\text{m}^2$$

$$\rho = 2695 \text{ kg}/\text{m}^3$$

$$v_0 = 3158 \text{ m}/\text{s}$$

$$\rho v_0^2 = 2.7 \times 10^{10} \text{ N}/\text{m}^2$$

Hence from Eq. 6.10b, $F_z = -0.51$. From Fig. 6.8 ($\eta = 0.5$), $P_z \simeq -0.75$. From Eq. 6.9b,

$$r = j\left(-0.75 \frac{K^2}{2} - 0.51 \frac{h}{\lambda}\right)$$

$$\simeq -j(0.51)\frac{h}{\lambda} \quad \left(\text{for } \frac{h}{\lambda} \sim 0.01\right)$$

since K^2 is small (0.11%) (Table 3.2).

(b) For a grooved array on ST-X quartz,

$$\alpha_x = 6.7 \times 10^{10} \text{ N}/\text{m}^2$$

$$\alpha_z = 8.7 \times 10^{10} \text{ N}/\text{m}^2$$

$$\rho = 2651 \text{ kg}/\text{m}^3$$

$$v_o = 3159 \text{ m/s}$$

$$\rho v_o^2 = 2.6 \times 10^{10} \text{ N/m}^2$$

From Eq. 6.10b, $F_z = -0.64$. Also, $P_z = 0$ for grooves. From Eq. 6.9b, $r = -0.j\,(0.64)\,(h/\lambda)$.

(c) For Y-Z lithium niobate,

$$-\frac{\pi K^2}{C_s}\,c_x^2 = 0$$

$$-\frac{\pi K^2}{C_s}\,c_y^2 = 1.1 \times 10^{-11} \text{ m}^2/\text{N}$$

$$-\frac{\pi K^2}{C_s}\,c_z^2 = -5.0 \times 10^{-11} \text{ m}^2/\text{N}$$

The constants for aluminum are the same as in part (a). Hence from Eq. 6.10b,

$$F_z = -0.24$$

From Fig. 6.8, $P_z \simeq -0.75$, as before.

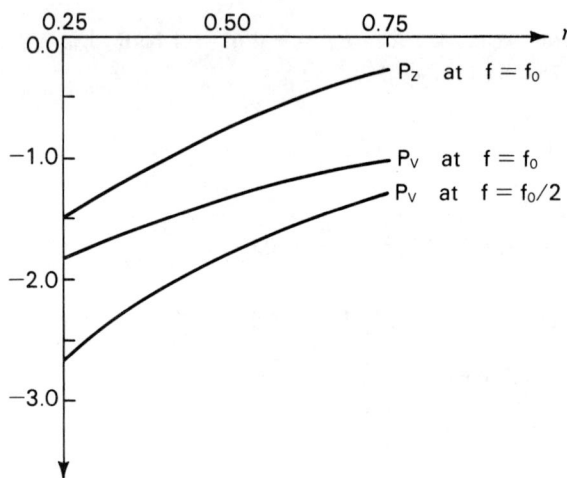

Figure 6.8 P_V and P_Z as functions of metallization ratio, η. P_v at $f = f_0/2$ is also included for use in the discussion of split-electrode transducers.

6 REFLECTORS

$$r = j\,-0.75\,\left(\frac{K^2}{2} - 0.24\,\frac{h}{\lambda}\right)$$

Usually, for $h/\lambda \sim 0.01$ the second term is negligible compared to the first terms because of the large K^2 of lithium niobate (see Table 3.2).

$$r \simeq -0.018j$$

This is the reflection coefficient per strip for *connected* electrodes on Y-Z lithium niobate. If the strips are left *unconnected*, a voltage appears which alternates in polarity from strip to strip just as in an IDT. This causes an additional regeneration reflection which can be calculated using the method discussed in Example 9.8 for IDTs. An incident SAW of amplitude ϕ^+ induced a voltage V_T across the electrodes given by

$$V_T = \frac{g_m \phi^+}{j\omega C_T}$$

assuming that $G_a \ll j\omega C_T$. This voltage regenerates a reflected wave of amplitude ϕ^- given by

$$\phi^- = \mu V_T = j\,\frac{G_a}{\omega C_T}\,\phi^+$$

where we have used the relation $G_a = -\mu g_m$. For an IDT with N pairs of electrodes having a metallization ratio $\eta = 0.5$,

$$\frac{G_a}{\omega C_T} = 1.28\,K^2 N \qquad \text{(see Example 9.1)}$$

For one electrode, $N = 0.5$, so that

$$\frac{\phi^-}{\phi^+} = 0.64jK^2 = 0.031j$$

The reflection coefficient for unconnected strips is equal to that for connected strips plus the regeneration reflection coefficient.

$$\begin{aligned}r_{\text{unconnected}} &= -0.018j + 0.031j \\ &= 0.013j\end{aligned}$$

Note that the reflection coefficient for unconnected strips is 180° out of phase with that for connected strips.

(d) For gold reflectors (Table 6.1),

$$\alpha_x = 2.85 \times 10^{10} \text{ N/m}2$$

$$\alpha_z = 9.8 \times 10^{10} \text{ N/m}^2$$

$$\rho = 19300 \text{ kg/m}^3$$

$$\rho v_0^2 = 19.25 \times 10^{10} \text{ N/m}2$$

The constants for the substrate are the same as in part (a). Hence from Eq. 6.10b,

$$r = j(1.94)\frac{h}{\lambda}$$

The reflectivities for different types of strips calculated in this example agree quite well in magnitude and phase with the experimental data in Ref. 6.5. The only exception is gold reflectors, for which the experimental value is 40% lower [$r = j(1.3)h/\lambda$]. The reason for the discrepancy is not clear; one possibility is that an evaporated thin film may have elastic constants that differ from those of the bulk material.

Example 6.5
Calculate the center-frequency shift of each of the reflectors in Example 6.4.

Solution
(a) and (b) From Eq. 6.10a, using the parameters in Example 6.4,

$$F_v = 0$$

From Fig. 6.8, $P_v \simeq -1.5$. From Eq. 6.9a,

$$\frac{v - v_0}{v_0} = -0.75 \frac{K^2}{2}$$

However, K^2 of ST-X quartz is very small, so that

$$\frac{v - v_0}{v_0} \simeq 0.$$

6 REFLECTORS

In this case the dominant contribution to $(v - v_o)$ comes from the stored energy effect, which we have not discussed at all. Clearly, our formulas are not very useful for this situation. However, for gold electrodes F_v will be nonnegligible and can give a dominant contribution. This is because of the large mass density (ρ) of gold.

(c) Y-Z lithium niobate has a large K^2, so that the dominant contribution to $(v - v_o)$ comes from P_v. From Fig. 6.8, $P_v = -1.4$, as before.

$$\frac{v - v_o}{v_o} = -0.75 \frac{K^2}{2}$$

For a fixed wavelength, frequency is directly proportional to the velocity.

$$\frac{\Delta f}{f_o} = -0.75 \frac{K^2}{2}$$

The center frequency is shifted downward from what we would expect using v_o from Table 3.2 $(f_o = v_o/2p)$. As in Example 6.4, this result is valid for connected strips; for unconnected strips it can be shown that regeneration causes an increase in velocity. At $\eta = 0.5$, the increase is $0.5K^2/2$, so that for unconnected strips at $\eta = 0.5$,

$$\frac{\Delta f}{f_o} = -0.25 \frac{K^2}{2}$$

6.3. Reflection of SAW by Electrodes: Similarities and Differences with Transmission Lines*

In this section we try to justify Eqs. 6.10 for the first-order mechanical effects F_z and F_v using heuristic arguments and point out the similarities and differences with transmission lines. In this book we have often used a transmission-line representation for SAW; however, in a SAW different modes are coupled together with many field components, and this gives rise to profound differences with the transmission line. Equation 6.10 for

---

*A modified version of the paper "Reflection and Mode...Media" by S. Datta appearing in 1983 ULTRASONICS SYMPOSIUM PROCEEDINGS, October 31 – November 2, 1983, Atlanta, GA, pp. 362-368. © 1983 IEEE.

F_z and F_v is one example of such a difference which we will now discuss. In a transmission line, we usually have $|F_z| = |F_v|$; that is, the impedance discontinuity is equal to the change in velocity. Consequently, we might expect that for SAW, too, an electrode that produces no velocity change should produce no reflections. This is not true, however, as we can easily see from a simple example. In the case of grooves there is no first-order change in velocity since the "electrodes" are the same as the substrate ($F_v = 0$); however, there is a first-order reflection, as we have seen ($F_z \neq 0$). This difference between F_z and F_v arises from the coupled-mode nature of the SAW.

The basic problem we wish to address is as follows. A surface wave of amplitude A^+ propagating on a substrate meets a discontinuity at the surface (Fig. 6.9). The discontinuity may be due to an electrode or due to

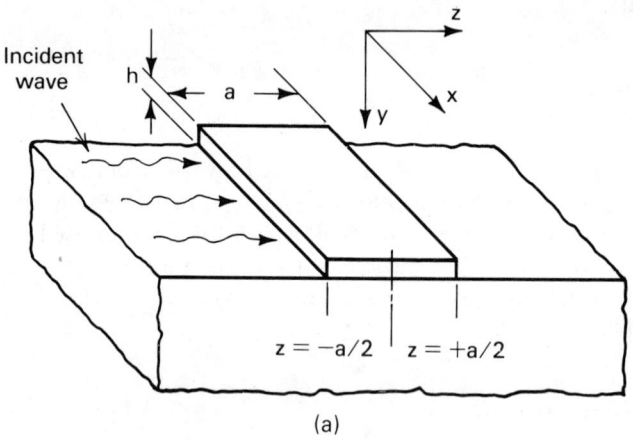

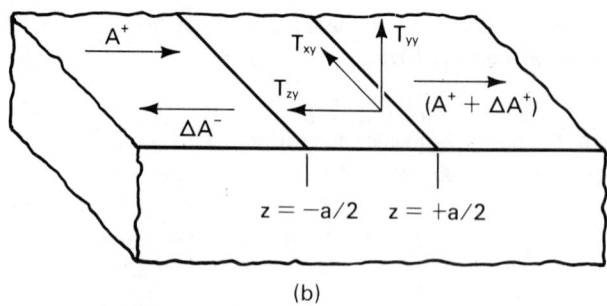

Figure 6.9 (a) Surface wave incident on thin-strip overlay; (b) generation of a reflected wave by stresses included at strip–substrate interface.

6 REFLECTORS

grooves etched on the surface. We wish to know the amplitude of the reflected wave, ΔA^-. This problem is in general a very difficult one; however, if the height h of the discontinuity is small compared to a wavelength ($h/\lambda < 0.01 - 0.02$), the problem can be treated in a fairly simple manner using perturbation theory. The condition of small h/λ is usually satisfied in practical devices; large values of h/λ result in significant bulk-mode conversion, which is unacceptable for device operation.

The method we will use is basically the same as that described by Auld in Ref. 1 (Vol. II, p. 302) with a modification to incorporate edge effects. A rigorous derivation of this theory based on the normal-mode analysis is beyond the scope of this book. Instead, we will develop it by analogy with a simple transmission line, pointing out the similarities and differences.

6.3.1. Transmission Lines

The simplest analogy to our present problem is that of reflections in a transmission line (Fig. 6.10). The discontinuity is represented by a region whose characteristic impedance $Z_0 + \Delta Z$ is slightly different from the usual impedance Z_0, with $\Delta Z/Z_0 \ll 1$. There is a reflection of $+\Delta Z/2Z_0$ at the leading edge and $-\Delta Z/2Z_0$ at the trailing edge. If we use the center of the discontinuity as our reference plane, the voltage reflection coefficient is given by

$$\frac{\Delta V^-}{V^+} = \frac{\Delta Z}{2Z_0}(e^{j\theta} - e^{-j\theta}) = j\frac{\Delta Z}{Z_0}\sin\theta \qquad (6.13)$$

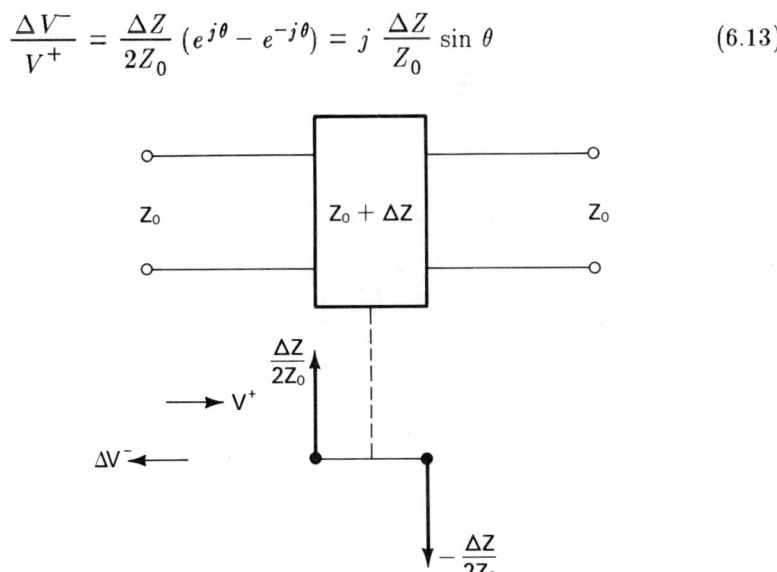

Figure 6.10 Reflection from a mismatched section in a transmission line.

where $\theta = ka = 2\pi a/\lambda$. Equation 6.13 gives us the reflection coefficient if we know $\Delta Z/Z_0$; the main problem in our case, of course, is to figure out the right $\Delta Z/Z_0$ for a given electrode-substrate combination. The same problem may be viewed in a somewhat different manner which gives the same results for transmission lines but generalizes more readily to our problem. The change in characteristic impedance ΔZ can be due to a change ΔL in the inductance L_0 per unit length and/or due to a change ΔC in the capacitance C_0 per unit length; let us assume that $\Delta L = 0$, so that

$$\frac{\Delta Z}{Z_0} = -\frac{\Delta C}{2C_0} \text{ since } Z_0 = \sqrt{\frac{L_0}{C_0}} \tag{6.14}$$

The incident wave V^+ induces a current $-j\omega \, \Delta C \, V^+$ when it meets the extra capacitance ΔC per unit length (Fig. 6.11). This current is given by

$$I(z) \, dz = -j\omega \, \Delta C \, dz \, V^+ \tag{6.15}$$

The induced current now acts as a source that generates reflected waves.

$$dV^- = \frac{-Z_0}{2} I(z) \, dz \tag{6.16}$$

Using Eqs. 6.14 and 6.15 and the relation $k = \omega C_0 Z_0$, we get

$$\frac{dV^-}{V^+} = -j(2k) \frac{\Delta Z}{2Z_0} \, dz \tag{6.17}$$

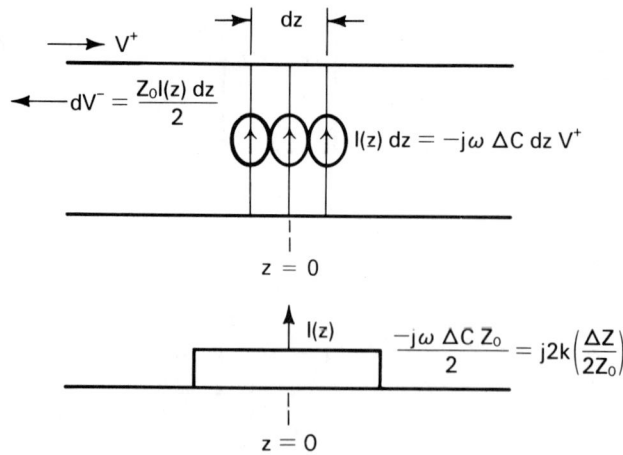

Figure 6.11 Current induced by incident wave at a mismatched section.

6 REFLECTORS

The total reflection is obtained by integrating Eq. 6.17 after introducing the phase factor e^{-j2kz} so as to shift all the reflections to a common reference plane.

$$\frac{\Delta V^-}{V^+} = \frac{\Delta Z}{2Z_0} \int_{-\theta}^{+\theta} e^{-j2kz} \, d(2kz) = j \frac{\Delta Z}{Z_0} \sin \theta \qquad (6.18)$$

as before. We note here that the discontinuity in the characteristic impedance is related to the induced current by the relation

$$\frac{\Delta Z}{2Z_0} = \frac{Z_0}{2} \frac{I}{2jkV^+} \qquad (6.19)$$

In our problem involving surface waves we will calculate the equivalent I/V^+ and then deduce an equivalent $\Delta Z/2Z_0$ using Eq. 6.19.

6.3.2. Reflection of SAW

Surface acoustic waves in piezoelectric solids have a fairly complicated pattern of acoustic and electrostatic fields coupled together. The first thing we need to figure out are the "voltages" and "currents." One way to do this is to look at the expression for power flow:

$$P_i = -\frac{1}{2}\left[v_x^* T_{xi} + v_y^* T_{yi} + v_z^* T_{zi} - \phi^*(j\omega D_i)\right], \quad i = x, y, z \quad (6.20a)$$

$$P = \frac{1}{2} V^* I \quad \text{(transmission line)} \qquad (6.20b)$$

Here v is the particle velocity, ϕ the electrostatic potential, D the electrical displacement, and T the stress. The stress is defined as force per unit area, and its two subscripts indicate the direction of the force and the direction of the outward normal to the area. For example, T_{xy} represents the x-directed force per unit area along the xz plane.

Comparing Eqs. 6.20a and 6.20b, it appears that a surface wave has four pairs of voltages and currents. Usually, v_x is zero or negligible and the particle motion is primarily in the sagittal plane (yz plane, Fig. 6.9); so we will ignore this component. The electrostatic part can be important if the discontinuity is produced by a conducting electrode and the material has a high piezoelectric constant. This effect, called *piezoelectric scattering*, gives rise to P_v and P_z which we will ignore. We concentrate on the mechanical reflection due to the component v_y (transverse) and v_z (longitudinal) of the particle motion. Let us look at these one by one.

Impedance discontinuity due to transverse component. First we consider the particle motion perpendicular to the surface. The induced current, I in this case, can be identified with $-T_{yy}$, which appears on the surface because of the presence of the strip (Fig. 6.9), while the voltage in this case is the particle displacement v_y^+ due to the incident wave. Hence the impedance discontinuity due to the transverse component can be written from Eq. 6.19 as

$$\left.\frac{\Delta Z}{2Z_0}\right|_y = \frac{-|v_y|^2}{4P_a} \frac{T_{yy}}{2jkv_y^+} \qquad (6.21)$$

where P_a is the power carried by the surface wave per unit beam width, and we have used $|v_y|^2/4P_a$ to replace $Z_0/2 \, (= V^2/4P)$. We now have to determine T_{yy}/v_y^+. This is done readily, assuming that the strip is a thin one so that it merely floats on the wave without loading it. From Newton's law (Fig. 6.12)

$$T_{yy} \, dz \, W = (\rho \, dz \, Wh) \frac{dv_y^+}{dt}$$

so that

$$\frac{T_{yy}}{v_y^+} = j\omega\rho h \qquad (6.22)$$

Using Eq. 6.22 in Eq. 6.21, we have

$$\left.\frac{\Delta Z}{2Z_0}\right|_y = \frac{-|v_y|^2}{4P_a} \frac{\rho v_0 h}{2} \qquad (6.23)$$

where $v_o = \omega/k$ is the velocity of the surface wave.

Impedance discontinuity due to longitudinal component. Arguing as before, we can write the impedance discontinuity due to the longitudinal component as

$$\left.\frac{\Delta Z}{2Z_0}\right|_z = \frac{-|v_z|^2}{4P_a} \frac{T_{zy}}{2jkv_z^+} \qquad (6.24)$$

Determining T_{zy}/v_z^+, however, is complicated by the fact that the forces on the side faces due to the stress T_{zz} have to be taken into account (Fig. 6.12). A similar component due to T_{yz} was ignored in Fig. 6.12a because

6 REFLECTORS

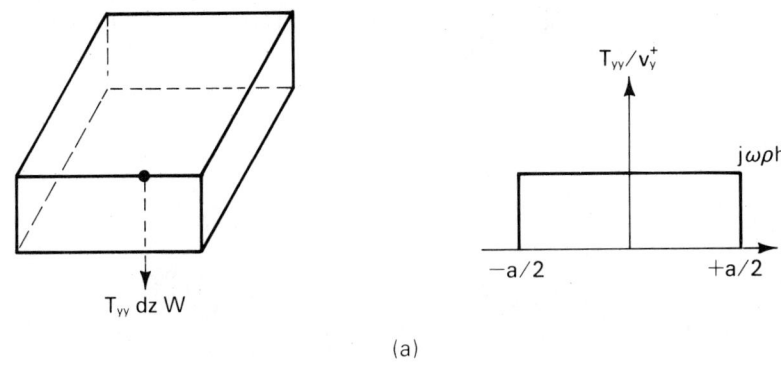

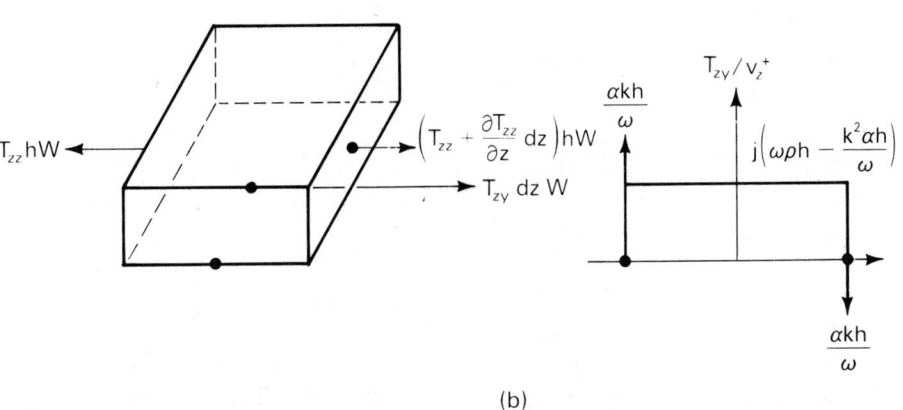

Figure 6.12 Stresses induced at strip–substrate interface by an incident surface wave: (a) transverse component; (b) longitudinal component.

T_{yz} is zero at the top surface of the strip, and its first-order value in the strip can be taken as zero. T_{zz}, on the other hand, is not zero at the top surface. From Newton's law (Fig. 6.12b),

$$T_{zy}\,dz\,W + \frac{\partial T_{zz}}{\partial z}\,Wh\,dz = (\rho\,dz\,Wh)\frac{\partial v_z^+}{\partial t}$$

that is,

$$\frac{T_{zy}}{v_z^+} = j\omega\rho h + jkh\,\frac{T_{zz}}{v_z^+} \tag{6.25}$$

where we have replaced $\partial/\partial z$ by $-jk$. If the strip material is isotropic with Lamé constants λ and μ, it can be shown that

$$\frac{T_{zz}}{v_z^+} = -\frac{\alpha k}{\omega} \tag{6.26}$$

where

$$\alpha = \frac{4\mu(\lambda + \mu)}{\lambda + 2\mu}$$

is an effective stiffness coefficient. If the strip material is not isotropic, the expression for α is more complicated. Using Eq. 6.26 in Eq. 6.25 yields

$$\frac{T_{zy}}{v_z^+} = j\omega\rho h - \frac{jk^2 \alpha h}{\omega} \tag{6.27}$$

Using Eq. 6.27 in Eq. 6.24 gives us

$$\left.\frac{\Delta Z}{2Z_0}\right|_z' = \frac{-|v_z|^2}{4P_a}\left(\frac{\rho v_0 h}{2} - \frac{\alpha h}{2v_0}\right) \tag{6.28'}$$

However, this result is not quite right. The reason is that in going from Eq. 6.25a to Eq. 6.25b, we replaced $\partial/\partial z$ by $-jk$. This is correct within the strip, but at the edges of the strip, T_{zz} must go to zero. If we assumed (as a first-order approximation) that T_{zz} is constant within the strip but goes to zero abruptly at the edges, the term $\partial T_{zz}/\partial_z$ also gives rise to delta-function stresses at the two edges of magnitude $(\alpha k h/\omega)v_z^+$, which we have neglected. These produce an additional impedance discontinuity of

$$\left.\frac{\Delta Z}{2Z_0}\right|_z'' = \frac{-|v_z|^2}{4P_a}\frac{\alpha h}{v_0} \tag{6.28''}$$

Adding Eqs. 6.28' and 6.28'' we get the total impedance discontinuity due to the longitudinal component.

$$\left.\frac{\Delta Z}{2Z_0}\right|_z = \frac{-|v_z|^2}{4P_a}\left(\frac{\rho v_0 h}{2} + \frac{\alpha h}{2v_0}\right) \tag{6.28}$$

6 REFLECTORS

Total impedance discontinuity. We can now obtain the total impedance discontinuity from the components $\Delta Z/2Z_0\big|_z$ (Eq. 6.23) and $\Delta Z/2Z_0\big|_z$ (Eq. 6.28). However, we have to remember that the phases of v_y and v_z are not related in the same way for the incident and reflected waves. For example, if $v_y \sim jv_z$ for the incident wave, then $v_y \sim -jv_z$ for the reflected wave. This can be understood easily if the crystal is reflection symmetric, that is, $+z$ looks the same as $-z$. In that case the fields of the reflected wave can be determined from those of the incident wave simply by turning the z axis around ($z \to -z$); it is clear that changing $z \to -z$ changes the sign of v_z but does not affect v_y. Hence $v_y^+/v_z^+ = -v_y^-/v_z^-$, where the superscripts $+$ and $-$ refer to incident and reflected waves, respectively.

Because of this change in sign, the net impedance discontinuity is actually the *difference* between $\Delta Z/2Z_0\big|_y$ and $\Delta Z/2Z_0\big|_z$.

$$\frac{\Delta Z}{2Z_0} = \frac{\Delta Z}{2Z_0}\bigg|_y - \frac{\Delta Z}{2Z_0}\bigg|_z \qquad (6.29)$$

The overall sign in Eq. 6.29 depends on whether we define the reflection coefficient as $\Delta v_y^-/v_y^+$ or as $\Delta v_z^-/v_z^+$. Usually, we are interested in the reflection coefficient $\Delta \phi^-/\phi^+$ because we like to place our transducers at the peak potential of the standing wave. Since ϕ, like v_y, is unaffected by the transformation $z \to -z$, we expect that $\Delta \phi^-/\phi^+ = \Delta v_y^-/v_y^+ = -\Delta v_z^-/v_z^+$. However, this is true only for reflection symmetric guides. To get the correct reflection coefficient for ϕ in all cases, we need a slight modification which is discussed at the end of this section. From Eqs. 6.23, 6.28, and 6.29,

$$\frac{\Delta Z}{2Z_0} = \frac{-|v_y|^2}{4P_a}\frac{\rho v_0 h}{2} + \frac{|v_z|^2}{4P_a}\left(\frac{\rho v_0 h}{2} + \frac{\alpha h}{2v_0}\right) \qquad (6.30)$$

The reflection coefficient from a strip of length a is given by

$$r = j\sin ka\left[\frac{|v_y|^2}{4P_a}\rho v_0 h - \frac{|v_z|^2}{4P_a}\left(\rho v_0 h + \frac{\alpha h}{v_0}\right)\right] \qquad (6.31)$$

At this point we should note a very important difference with a transmission line. For a transmission line $Z_0 = \sqrt{L/C}$ and $v_0 = \sqrt{1/LC}$, so that if we change C keeping L constant, we expect that

$$\frac{\Delta Z}{2Z_0} = \frac{\Delta v}{2v_0}$$

However, in the case of the surface wave this is *not* true. First, the delta-function stresses that we encountered have no effect on velocity; they are equal and opposite at the edge of the strip and cancel each other out. For this reason we should use Eq. 6.28' rather than Eq. 6.28. Second, the transverse and longitudinal components *add* rather than subtract, so that

$$2\pi \frac{\Delta v}{2v_0} = \frac{-|v_y|^2}{4P_a} \frac{\rho v_0 h}{2} - \frac{|v_z|^2}{4P_a} \left(\frac{\rho v_0 h}{2} - \frac{\alpha h}{2v_0}\right) \qquad (6.32)$$

Equation 6.32 has the simple interpretation that the mass loading (ρ) tends to slow the wave down while the stiffness (α) tends to speed it up. Comparing with Eq. 6.31, it is apparent that although the same terms are involved, no such simple interpretation is possible. Moreover, it is clear that discontinuities that produce minimal velocity perturbation do *not* produce minimal reflection, and vice versa. Equations 6.30 and 6.32 can be put in the form of Eq. 6.10 using Eq. 3.22 and the definitions of $c_{y,z}$.

$$\frac{\Delta Z}{Z_0} = \frac{h}{\lambda} \frac{\pi K^2}{C_s} \left[-|c_y|^2 \rho v_0^2 + |c_z|^2 (\alpha_z + \rho v_0^2)\right] \qquad (6.33)$$

$$\frac{\Delta v}{v_0} = \frac{h}{\lambda} \frac{K^2}{2C_s} \left[-|c_y|^2 \rho v_0^2 + |c_z|^2 (\alpha_z - \rho v_0^2)\right] \qquad (6.34a)$$

Actually (as we mentioned at the beginning of this section), Eq. 6.33 gives the correct sign of the reflection coefficient for v_y, which is the same as that for ϕ only if c_y is pure imaginary (see Ref. 0.1, Vol. II, p. 180, for an explanation). This is true of crystals that are reflection-symmetric in z; however, as we can see from Table 3.1, many substrates do not satisfy this condition and we have to modify Eq. 6.33 slightly in order that it yield the correct sign of the reflection coefficient for ϕ (and not v_y) in all cases.

$$\frac{\Delta Z}{Z_0} = \frac{h}{\lambda} \frac{\pi K^2}{C_s} \left[-c_y^2 \rho v_0^2 - c_z^2(\alpha_z + \rho v_0^2)\right] \qquad (6.34b)$$

Equations 6.34a and 6.34b are the same as Eqs. 6.10a and 6.10b, which we stated without proof in Section 6.2. The only difference arises because we neglected c_x for clarity in our present discussion.

7
ATTENUATORS AND AMPLIFIERS

A thin conductive layer at the surface produces a reduction in the velocity of the surface wave (Section 3.3). If the layer is not perfectly conducting, it will produce an attenuation, whereas if we apply a dc electric field of sufficient magnitude to the resistive layer we can get an amplification. In Section 7.1 we discuss the amplification and attenuation caused by the electrical loading of thin layers. The mechanical loading due to thick layers and fluids is taken up in Section 7.2.

7.1. Electrical Loading

In Section 3.3 we saw that a thin conducting layer (see Fig. 3.4) produces a change in velocity of the surface wave by acting as a capacitive load ΔC on the SAW transmission line (see Eqs. 3.5, 3.12, and 3.13).

$$\frac{\Delta v_o}{v_o} = -\frac{K^2}{2} = -\frac{\Delta C}{2} Z_o v_o \qquad (7.1\text{a})$$

The capacitance ΔC is determined from the charge density ρ_s induced in the layer by the SAW whose amplitude is ϕ (Eq. 3.14).

$$\Delta C = -\frac{\rho_s W}{\phi} \qquad (7.1\text{b})$$

The fractional change in velocity ($\Delta v/v$) is the negative of the fractional change in wave number ($\Delta k/k$). Using this fact in Eq. 7.1a and combining with Eq. 7.1b, we have

$$\frac{\Delta k}{k} = \frac{K^2}{2} = \frac{Z_o v_o W}{2} \left(\frac{-\rho_s}{\phi}\right) \tag{7.2}$$

As we saw in Section 3.3, for the conducting layer, ρ_s is 180° out of phase with ϕ, giving a positive Δk (and a negative Δv). What happens if we have a layer whose electrical properties are such that ρ_s/ϕ is a complex number? Δk will then have a real part β and an imaginary part α.

$$\Delta k = \beta + j\alpha \tag{7.3}$$

We will now have fields that vary as $e^{-j(k+\beta+j\alpha)z}$, implying attenuation (negative α) or amplification (positive α). The loading is no longer purely capacitive but has a resistive component; the resistance is positive for attenuation and negative for amplification. The nature of the loading is basically determined from the charge induced in the layer by the surface wave. We may write

$$\frac{\beta + j\alpha}{k} = \frac{K^2}{2} \frac{\rho_s'}{\rho_s} \tag{7.4}$$

where ρ_s' is the charge density induced in the layer by a SAW of amplitude ϕ and ρ_s is the charge density that would be induced in a perfectly conducting thin layer by the same SAW. ρ_s is related to ϕ by

$$\rho_s = -kC_s \phi \tag{7.5}$$

It can easily be checked that Eq. 7.5 inserted in Eq. 7.2 gives the relation $K^2 y_o = 2\pi C_s v_o$.

Now let us consider a thin layer whose thickness is d and conductivity is σ. We need to find the charge density ρ_s' induced by a surface wave of amplitude ϕ; this can be done exactly, but to gain some insight let us look at a limiting case. Suppose the conductivity σ is so low that ρ_s' is very small and the potential in the layer is nearly equal to ϕ. This is the opposite of the highly conducting layer, where the induced charge ρ_s' produced a potential, ψ_s, big enough to cancel out ϕ (see Eq. 3.20). The longitudinal electric field E_z in the layer is then equal to $jk\phi$, producing a current density, $J_z = j\sigma k\phi$. But from the continuity equation,

7 ATTENUATORS AND AMPLIFIERS

$$\frac{\partial J_z}{\partial z} = -\frac{\partial}{\partial t}\frac{\rho_s'}{d}$$

We have divided ρ_s' by d to get the bulk charge density, assuming that it is uniform across the thickness d. We then have

$$\rho_s' = \frac{kd}{\omega} J_z$$

$$= j\frac{k\sigma_\square}{v_o}\phi \tag{7.6}$$

where $\sigma_\square = \sigma d$ is the sheet conductivity and $v_o = \omega/k$.

We now have two limiting results: Eq. 7.5 for highly conductive layers and Eq. 7.6 for highly resistive layers. For layers of intermediate conductivity it can be shown that

$$\rho_s' = -\frac{kC_s}{1 + (jv_o C_s/\sigma_\square)}\phi = \frac{\rho_s}{1 + (jv_o C_s/\sigma_\square)} \tag{7.7}$$

This is not meant to be obvious, but it is easily seen that Eq. 7.7 goes over to Eqs. 7.5 and 7.6 in the appropriate limits. Using Eq. 7.7 in Eq. 7.4, we have the attenuation and velocity change of the SAW caused by a layer of arbitrary conductivity.

$$\frac{\beta + j\alpha}{k} = \frac{K^2}{2}\frac{1}{1 + (jv_o C_s/\sigma_\square)} \tag{7.8a}$$

Hence

$$\frac{\beta}{k} = \frac{K^2}{2}\frac{1}{1 + (v_o C_s/\sigma_\square)^2} \tag{7.8b}$$

$$\frac{\alpha}{k} = -\frac{K^2}{2}\frac{v_o C_s/\sigma_\square}{1 + (v_o C_s/\sigma_\square)^2} \tag{7.8c}$$

Figure 7.1 shows β/k and α/k for changing values of $v_o C_s/\sigma_\square$. The fractional velocity change is the negative of β/k. As we may expect, it tends to $-K^2/2$ for high conductivity ($v_o C_s/\sigma_\square \ll 1$) and to zero for low conductivity ($v_o C_s/\sigma_\square \gg 1$). At $v_o C_s = \sigma_\square$, the fractional velocity change is halfway between the short circuit and the open circuit and the attenuation is a maximum. Thus the quantity $v_o C_s/\sigma_\square$ is a measure of what is a good conductor and what is not.

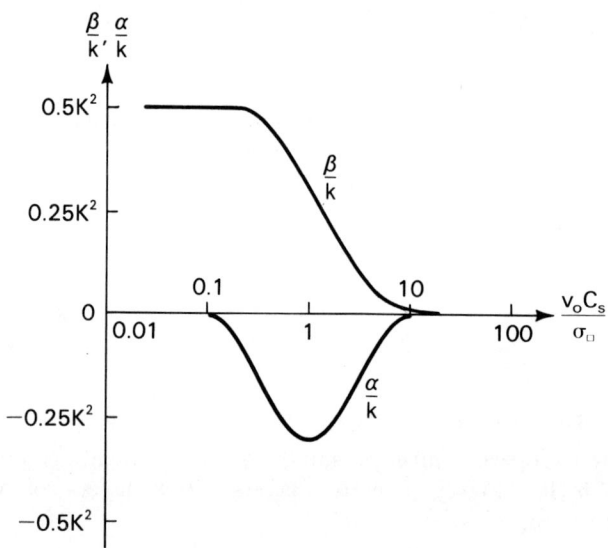

Figure 7.1 Phase shift and attenuation as a function of conductivity.

Example 7.1
Calculate the sheet resistivity for a layer to produce the maximum attenuation in Y-Z lithium niobate. What is the maximum attenuation?

Solution

$$v_o = 3500 \text{ m/s} \quad \text{(from Table 3.2)}$$

$$C_s = 4.6 \text{ Pf/cm}$$

Sheet resistivity for maximum attenuation:

$$\rho = \frac{1}{v_o C_s} = 621 \text{ k}\Omega/\text{square}$$

$$\frac{\alpha_{max}}{k} = -\frac{K^2}{4} = -0.012$$

$$\alpha_{max} = -0.012k = -0.65 \text{ dB/wavelength}$$

In general, the maximum possible attenuation due to electrical loading is $13.7K^2$ dB per wavelength.

7 ATTENUATORS AND AMPLIFIERS

Now let us put a battery of voltage V across the semiconductor layer (Fig. 7.2). This produces a dc electric field E which we can calculate by dividing V by the length of the layer. This electric field causes charge carriers in the semiconductor to drift with a velocity v_d given by

$$v_d = \mu E$$

where μ is the mobility of the carriers. Let us assume positive charge carriers (holes) so that they drift in the direction of the field, rather than against it. Without the battery, we have seen that the semiconductor causes an attenuation of the SAW because the SAW drags the charge carriers along with it, losing energy in the process. Actually, a charge carrier is alternately accelerated and retarded by a traveling wave. But it spends more time moving with the wave (because of the lower relative velocity) than against it, so that the net effect is to transfer energy from the wave. With the battery present, however, the charge carriers are already in motion with a velocity v_d. If $v_d = v_o$, we expect the charges to move with the wave anyway, so that the SAW does not do any work on the charges. So the attenuation α should go to zero. If we increase the battery voltage so that $v_d > v_o$, we have a reversal of roles. The charges want to move faster than the wave and tend to give up energy to the wave; the wave is amplified rather than attenuated.

The drift velocity v_d is easily incorporated mathematically by replacing v_o in Eq. 7.8 with $v_o - v_d$. We will not prove this, but it does seem reasonable to replace the wave velocity, v_o by the wave velocity $(v_o - v_d)$ relative to the electrons.

$$\frac{\beta}{k} = \frac{K^2}{2} \frac{1}{1 + \left[C_s(v_o - v_d)/\sigma_\square\right]^2} \tag{7.9a}$$

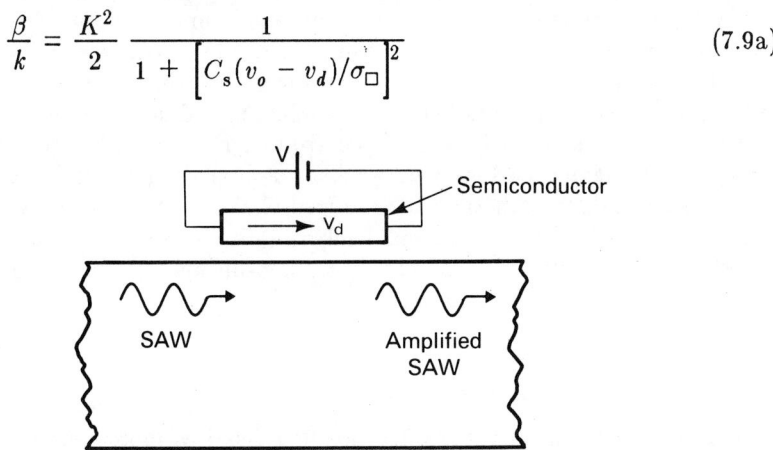

Figure 7.2 Surface-wave amplifier.

$$\frac{\alpha}{k} = -\frac{K^2}{2} \frac{C_s(v_o - v_d)/\sigma_\Box}{1 + \left[C_s(v_o - v_d)/\sigma_\Box\right]^2} \qquad (7.9b)$$

Example 7.2
What is the maximum amplification in a SAW amplifier, and under what conditions is it achieved? Discuss.

Solution
For amplification $v_d > v_o$ so that α is positive Eq. 7.9b. The maximum value is obtained for α when

$$C_s(v_d - v_o) = \sigma_\Box \quad \text{that is, when} \quad v_d = v_o + \frac{\sigma_\Box}{C_s}$$

Also,

$$\alpha_{max} = \frac{K^2}{4} k = 13.7 K^2 \text{ dB/wavelength}$$

As we can see, the required drift velocity v_d can be reduced to v_o if the conductivity $\sigma_\Box$ is small compared to $v_o C_s$. However, the small conductivity should not be the result of a small mobility μ; the field required to produce v_d will then be too high, causing excessive losses. For best results what is needed is a thin layer with high mobility and few carriers. The performance of surface wave amplifiers has been severely limited by surface state problems.

We should mention that we have simplified our discussion of amplifiers considerably by assuming a very thin layer. The thickness of the layer should be small compared not only to a wavelength, but also to what is called the Debye length. Otherwise, significant modifications in Eq. 7.7 are expected which have to be calculated from a more detailed treatment of the carrier dynamics. Moreover, we have assumed that the surface wave potential is small enough that the induced charge density modulation ρ_s' is much smaller than the constant background density of carriers. With large wave amplitudes we enter a non-linear regime where both attenuation and amplification saturate.

7.2. Mechanical Loading

A detailed discussion of mechanical loading requires more familiarity with acoustic quantities such as stress and particle velocities. For this reason we will merely state the mechanical effects rather than derive them. A

7 ATTENUATORS AND AMPLIFIERS

thin layer of solid at the surface produces a change in the velocity of the surface wave much like the thin conducting layer. The fractional change in wave number (which is the negative of the fractional change in velocity) is given by

$$\frac{\beta}{k} = \frac{K^2}{2C_s} \left[|c_x|^2 (\rho v_o^2 - \alpha_x) + |c_y|^2 \rho v_o^2 + |c_z|^2 (\rho v_o^2 - \dot{\alpha}_z) \right] \quad (7.10)$$

where K^2, C_s, and $c_{x,y,z}$ are properties of the substrate listed in Tables 3.1 and 3.2, while ρ, α_x, and α_z are properties of the layer as defined in Section 6.2. ρ is the density and α_x and α_z are effective stiffness coefficients. No attenuation is produced.

A fluid on the surface produces attenuation. For example, it is well known that a drop of water at the surface is very effective in absorbing the surface wave. This is because fluids usually have a large viscosity, which is like having an imaginary component to the stiffness coefficients. This gives an attenuation, as we can see from Eq. 7.10 if we let α_x and α_z have imaginary parts.

There is another distinct mechanism for attenuation. It is well known that at high frequencies the presence of air above the surface causes significant attenuation of the surface wave, which can be eliminated by evacuating the package. Here the loss is due to the excitation of compressional waves in air, which carry power away. The attenuation is given by

$$\frac{\alpha}{k} = \frac{K^2}{2C_s} |c_y|^2 (\rho v_o^2) \frac{v_c}{\sqrt{v_o^2 - v_c^2}} \quad (7.11)$$

where ρ is the density of air and v_c is the compressional wave velocity in air.

Example 7.3
Calculate the attenuation due to air loading in Y-Z lithium niobate.

Solution

$$\rho(air) = 1.21 \text{ kg/m}^3$$

$$v_c = 343 \text{ m/s}$$

For Y-Z lithium niobate, from Tables 3.1 and 3.2,

$$\frac{K^2}{2C_s}|c_y|^2 = 1.69 \times 10^{-12} \text{ m}^2/\text{N}$$

$$\frac{\alpha}{k} = 2.85 \times 10^{-6} \qquad (7.11)$$

$$\alpha = 1.6 \times 10^{-4} \text{ dB/wavelength}$$

At

$$100 \text{ MHz}, \alpha = 0.046 \text{ dB/cm}$$

$$1 \text{ GHz}, \alpha = 0.46 \text{ dB/cm}$$

The actual experimentally measured values are about 20% higher.

8
WAVEGUIDES

A surface wave beam of finite width diffracts as it propagates; the angle of diffraction is approximately given in radians by $(W/\lambda)^{-1}$, where W is the aperture width. For narrow beams this can prove to be a serious problem. In convolvers, for example, narrow beams ($<5\lambda$) are used to get higher power density and hence bigger nonlinear effects. At the same time long propagation lengths are desirable. It thus becomes important to use a waveguide.

In this chapter we discuss the basic principles for designing SAW waveguides. A well-known principle in wave propagation is that if a region with low wave velocity is surrounded by faster regions, the wave is confined to the slower region; so to make a waveguide what we need is a region with a lower SAW velocity. The easiest way to do this is with a thin metallic overlay (Fig. 8.1) which lowers the velocity. Alternatively, thick overlays of materials that increase the velocity can be used in the surrounding regions, with the wave guided in the slot between them.

8.1. Isotropic Substrate

Most substrates used for SAW devices are anisotropic and this has to be taken into consideration in the design of waveguides. However, let us first discuss the behavior of waveguides on an isotropic substrate; we will then consider the modifications introduced by anisotropy in the next section.

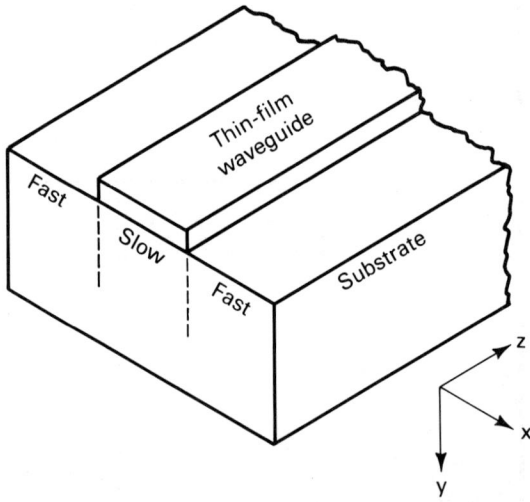

Figure 8.1 Thin-film SAW waveguide.

Let us consider a region between $x = -a/2$ and $x = +a/2$ where the wave velocity (v_s) is lower than in the surrounding regions (v_f) (Fig. 8.2). The wave is guided along the slow region with an amplitude profile as shown. Note that the wave extends somewhat outside the guiding region. This is in contrast to the metallic waveguides used for microwaves which confine the wave completely within the guide. The modes for such microwave waveguides are readily obtained by noting that the mth mode has an amplitude profile $\phi(x)$ given by

$$\text{Symmetric:} \quad \phi(x) = \phi(0) \cos \frac{m\pi x}{a} \quad m \text{ odd} \quad (8.1a)$$

$$\text{Antisymmetric:} \quad \phi(x) = \phi(0) \sin \frac{m\pi x}{a} \quad m \text{ even} \quad (8.1b)$$

Note that it goes to zero at the edges of the guide, $x = \pm a/2$. The wave number of the guided wave k_g is related to that of the unguided wave k by

$$k_g^2 = k^2 - \left(\frac{m\pi}{a^2}\right) \tag{8.2}$$

In the surface-wave waveguide, however, the wave hangs out beyond the guide and the amplitude profile in the guide can be written as

$$\text{Symmetric:} \quad \phi(x) = \phi(0) \cos \frac{m\pi x}{a + \delta_m} \quad m \text{ odd} \quad (8.3a)$$

8 WAVEGUIDES

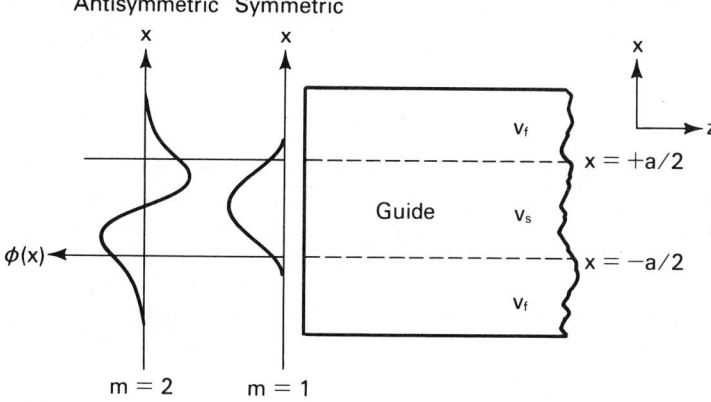

Figure 8.2 Amplitude profiles of the two lowest-order guided models.

Antisymmetric: $\phi(x) = \phi(0) \sin \dfrac{m\pi x}{a + \delta_m}$ m even (8.3b)

$$-\frac{a}{2} < x < +\frac{a}{2}$$

where δ_m represents an effective widening of the guide which is different for different modes. As we may expect, we now have

$$k_g^2 = k^2 - \left[\frac{m\pi}{a + \delta_m}\right]^2 \tag{8.4}$$

This is a common result in the theory of dielectric waveguides which are used in millimeter waves and integrated optics.

We need to know δ_m in order to use Eq. 8.4. The value of δ_m is determined by the nature of the boundary conditions between the guide and the surroundings. We will not go into the details, but if we require that $\phi(x)$ and $d\phi(x)/dx$ be continuous at the boundaries, it can be shown that (Refs. 8.1 and 8.2)

$$\delta_m = a \frac{2\theta_m}{m\pi - 2\theta_m} \tag{8.5a}$$

where

$$\sin \theta_m = \frac{m\pi}{k(a + \delta_m)\sqrt{1 - v_s^2/v_f^2}} \tag{8.5b}$$

Equations 8.5a and 8.5b have to be solved simultaneously for θ_m and δ_m. The wave number of the guided wave k_g is then obtained from Eq. 8.4 and the amplitude profile from Eq. 8.3.

Example 8.1
Consider an isotropic substrate with a region 4λ wide where the velocity is 2% slower than in the surroundings. Calculate the number of allowed modes and their phase velocities relative to that of the unguided mode.

Solution

$$a = 4\lambda \qquad ka = 8\pi$$

$$1 - \frac{v_s^2}{v_f^2} = 0.2$$

From Eqs. 8.5a and 8.5b,

$$\frac{\delta_m}{a} = \frac{2\theta_m}{m\pi - 2\theta_m} \qquad (8.6a)$$

$$\sin\theta_m = \frac{m\pi - 2\theta_m}{ka\sqrt{1 - v_s^2/v_f^2}} = \frac{m\pi - 2\theta_m}{1.6\pi} \qquad (8.6b)$$

One way to solve Eq. 8.6b is to plot both sides against θ. As we can see from Fig. 8.3, there are two modes $m = 1$ (symmetric) and $m = 2$ (antisymmetric):

$$\text{Symmetric, } m = 1: \quad \theta_1 = 26° = 0.14\pi$$
$$\frac{\delta_1}{a} = 0.39$$
$$\text{Antisymmetric, } m = 2: \quad \theta_2 = 58° = 0.32\pi$$
$$\frac{\delta_2}{a} = 0.47$$

Note that higher-order modes extend more outside the guide region, as shown by the higher δ/a. From Eqs. 8.4 and 8.5b,

$$\left(\frac{k_g}{k}\right)^2 = 1 - \left[\left(-\frac{v_s^2}{v_f^2}\right)\sin^2\theta_m\right]$$

8 WAVEGUIDES

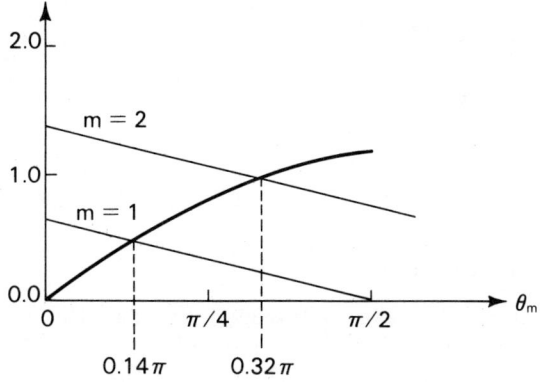

Figure 8.3 Solution of Equation 8.6b.

$$\frac{k_g}{k} = 0.996 \rightarrow v = 1.004 v_s \quad \text{(symmetric, } m = 1\text{)}$$

$$\frac{k_g}{k} = 0.985 \rightarrow v = 1.015 v_s \quad \text{(antisymmetric, } m = 2\text{)}$$

Example 8.2
Plot v/v_s for the two lowest modes as a function of ka.

Solution
We have to go through the procedure in Example 8.1 for different values of ka. The result is shown in Fig. 8.4. As we can see, the upper limit to v is $1.02 v_s$, which is the velocity in the fast region, while for wide guides v_g tends to v_o. The values of θ, δ/a, and v/v_s are given below for four values of ka.

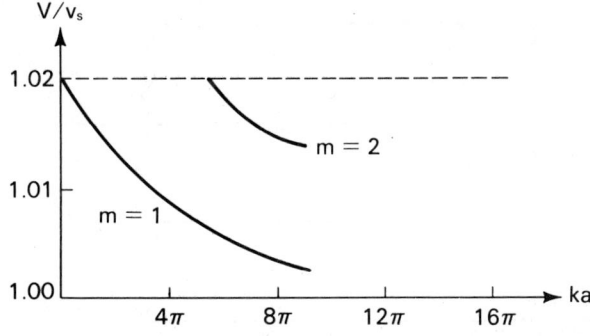

Figure 8.4 Dispersion curves for the two lowest modes ($m = 1, 2$).

ka	m = 1 θ	δ/2	v/v_s	m = 2 θ	δ/a	v/v_s
2π	0.32π	1.76	1.015	—	—	—
4π	0.23π	0.85	1.009	—	—	—
8π	0.14π	0.39	1.004	0.32π	0.47	1.015

8.2. Anisotropic Substrate

Most practical substrates are anisotropic, so that the velocity of the surface wave changes with direction. This affects the guide properties in the following way. As we have seen, the wave in the guide travels in the z direction but forms a standing wave in the x direction. The wave can be viewed as the superposition of two zigzag waves that travel at a slight angle, α to the z axis (Fig. 8.5). Thus we should use the velocity $v(\alpha)$ rather than the velocity in the z direction. Of course, $v(\alpha) = v(-\alpha)$ for pure mode directions (see Section 3.4). This description of a guided wave as a superposition of two plane waves propagating at an angle is familiar in the theory of waveguides for electromagnetic waves. However, reflection of the wave at the interface is similar to that in dielectric waveguides rather than the more familiar metallic waveguides; if the zigzag angle exceeds a certain critical value, there is no total reflection at the interface and guided waves cannot exist.

It can be shown that the zigzag angle α is given by

$$\tan \alpha = \frac{m\pi - 2\theta_m}{ka} \tag{8.7}$$

To account for anisotropy we should use $k(\alpha)$ instead of k in Eqs. 8.4 and 8.5b. but the main correction comes from using $v_s(\alpha)$ instead of v_s in Eq. 8.5b, since v_s/v_f is very close to 1.

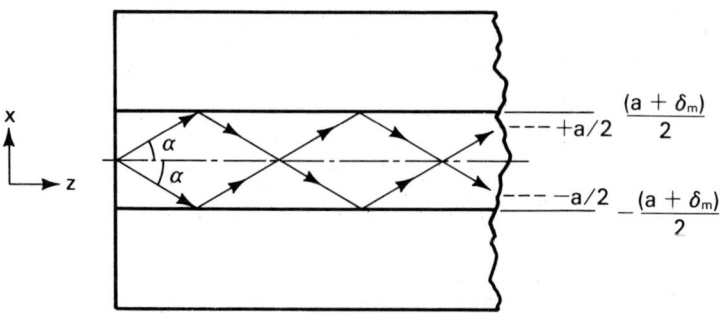

Figure 8.5 Zigzag waves in a waveguide.

8 WAVEGUIDES

Example 8.3
Calculate the change in the phase velocity plot (Fig. 8.4) for the lowest mode in Example 8.2, considering an anisotropic substrate in which the wave number at an angle α from the z axis is given by (α in radians)

$$k(\alpha) = k(0)\left(1 + \frac{\alpha^2}{4}\right)$$

Solution
Using θ from Example 8.2, we can calculate α from Eq. 8.7. This can be used to modify the value of $\sqrt{1 - v_s^2/v_f^2}$ and recalculate θ from Eq. 8.6b; v/v_s can then be determined as before. Usually, the correction is small so that a further iteration is not necessary.

ka	α	$\sqrt{1 - v_s^2/v_f^2}$
2π	10.2	0.235
4π	7.69°	0.221
8π	6.14°	0.213

9 | Part 3: SAW Devices

BANDPASS FILTERS

In Chapters 4 through 8 we have discussed individual components used in SAW devices. In the next two chapters we describe two types of SAW devices which have found widespread application in signal processing. Our discussion up to this point has been in terms of admittance functions, transfer functions, and so on, that characterize the device components; in the following chapters we will concentrate on overall device parameters, such as insertion loss, that are of interest to system designers.

A bandpass filter typically consists of an input transducer and an output transducer with a delay or propagation region in between (Fig. 9.1); if both transducers are apodized, a multistrip coupler is interposed between the transducers. The characteristics of these components have been discussed individually in Chapters 4 and 5. In this chapter we discuss the overall device performance using some of the results derived earlier.

In Section 9.1 we discuss the kinds of frequency response specifications that can be achieved with SAW bandpass filters. The basic design procedure is outlined in Section 9.2. Section 9.3 describes the analysis of the transmitting IDT and the receiving IDT. Starting from the equivalent-circuit model of Chapter 4, we discuss how the insertion loss and reflection can be determined. In Section 9.4 we discuss the various second-order effects that degrade filter performance and the techniques used to minimize or correct them. We conclude this chapter with two illustrative examples of filter design kindly provided by the Andersen Laboratories and the Phonon Corporation (Section 9.5).

A review of Chapter 1 may be useful before starting this chapter.

9 BANDPASS FILTERS

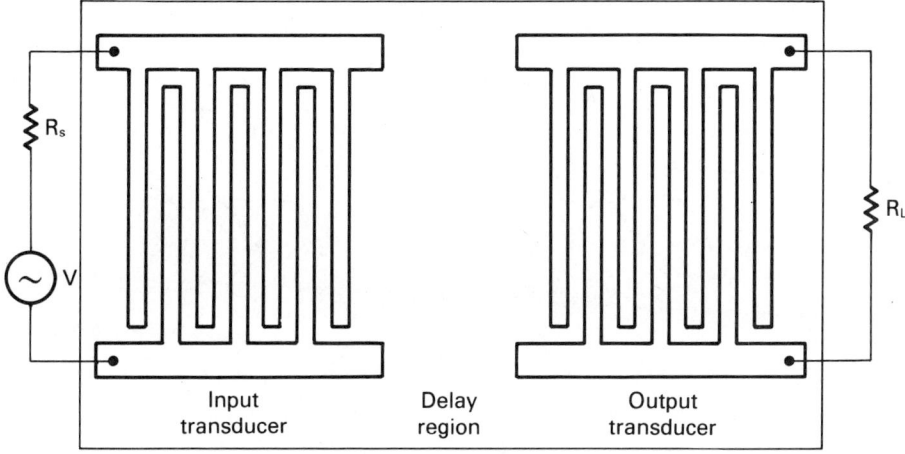

Figure 9.1 SAW bandpass filter.

9.1. Capabilities of SAW Bandpass Filters

TABLE 9.1: Practical Parameter Values for SAW Bandpass Filters

Center frequency, f_c	10–1000 MHz
1-dB fractional bandwidth, $\Delta f/f_c$	<40%
Transition bandwidth, f_T	>0.35 MHz
Shape factor, SF	>1.04
Amplitude ripple, A_r	>0.2 dB
Out-of-band rejection, R	<60 dB
Insertion loss, IL	>6 dB
Phase deviation from linearity	±3°
Power handling capability	<30 dbm

Figure 9.2 shows some of the typical parameters used to specify bandpass filter responses. Table 9.1 lists the practical range of values of these parameters that can be attained with SAW devices. Manufacturing tolerance and temperature variations also limit the transition bandwidth since the designer must "overdesign" the part to meet the requirements over an operational temperature range and with errors introduced in manufacturing.

Before proceeding to the detailed design and analysis of devices, let us briefly discuss the factors that control the parameters in Table 9.1. As we have seen in Chapter 1, a SAW filter is basically a tapped delay line in which the tap weights replicate the sampled impulse response of the bandpass filter; solid-electrode transducers use two samples per cycle and

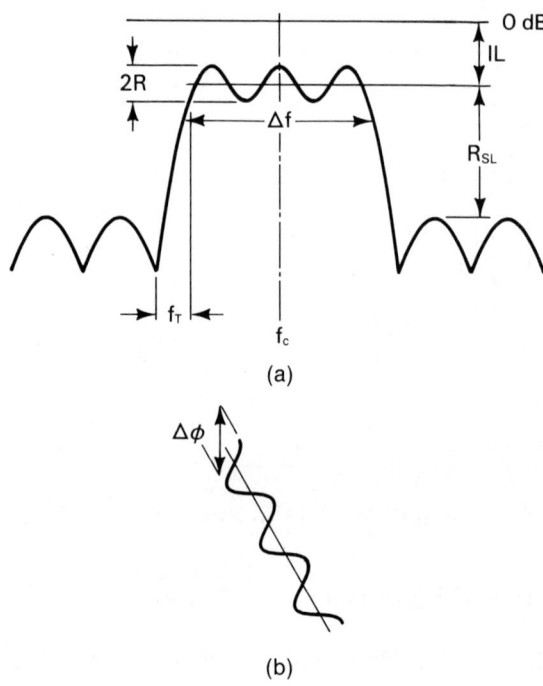

Figure 9.2 Frequency-response specifications for bandpass filters: (a) magnitude response; (b) phase response.

split-electrode transducers use four samples per cycle to minimize internal reflections. SAW filter designs are thus typically done in the time domain and the limitations on different parameters are best understood from the impulse response. For example, the upper limit of the *center frequency* f_c is imposed by photolithographic capabilities since the samples get too close together at high frequencies; at 1000 MHz, assuming a typical SAW velocity of 3000 m/s, the center-to-center separation of (solid) electrodes is 1.5 μm. This requires 0.75-μm-wide electrodes, which represents the limit of present-day photolithography. Higher-frequency devices can be built using electron-beam lithography; the high-frequency limitation then arises from losses in the substrate and due to air loading. On the low-frequency side the limitation is basically due to the large size and thickness of the substrate required.

The *transition bandwidth* (f_T) is limited by the length of the transducer that can be practically synthesized. The time duration of the impulse response is approximately twice the reciprocal f_T (the precise value depends on the passband ripple; see Section 1.4 on window functions); an f_T of 0.35 MHz thus requires an impulse response ~6 μs long. Assuming a SAW velocity of 3000 m/s, this requires a transducer 1.8

9 BANDPASS FILTERS

cm long. The *insertion loss* (*IL*) is usually greater than 6 dB (3 dB per transducer) because the transducers are bidirectional; often we have to settle for higher insertion loss to reduce spurious echoes due to reflections that cause undesirable amplitude and phase ripples in the passband. In future, the development of convenient unidirectional transducers may lead to lower insertion loss. These points are discussed more fully in Section 9.3. There are two factors that cause *amplitude ripples (R)* in the passband. It might be inherent in the design due to the window function used (Section 1.4), or it may arise from the triple transit echo, which is discussed in Section 9.4.3. The triple transit echo also causes ripples in the phase response, making it deviate from the linear response expected from an ideal symmetric impulse response. Some phase ripple also arises from the radiation susceptance, B_a (Section 9.3.1).

Large *fractional bandwidths* ($\Delta f / f_c$) require a short transducer with few electrodes; this leads to increased IL unless a strongly piezoelectric substrate is used. Thus wideband filters are usually built on lithium niobate, which has a large piezoelectric coupling constant. The *out-of-band rejection* (R_{SL}) is usually determined by the accuracy with which tap weights can be implemented. This is because the degree of rejection depends on the precision with which the signal from different electrodes cancel each other. The tap weight precision is limited by second-order effects such as diffraction and end effects. Generation of spurious bulk waves and plate modes also reduces the rejection level on the high-frequency side of the passband (Section 9.4). The *power-handling capability* is limited by breakdown due to the voltage between the electrodes; SAW devices are passive, linear devices and nonlinearities do not impose any practical limit on the power level.

9.2. Basic Design Procedure

In designing a SAW device to meet a particular specification, the first step is to decide on a substrate material; typical choices are Y-Z lithium niobate for wideband filters and ST quartz for narrowband filters. The overall frequency response of a device is the product of the responses of the input and output transducers. A designer has to decide on how the desired frequency response can be divided between the two. It is difficult to achieve more than 30 dB of out-of-band rejection from one transducer, so that if a 50-dB rejection level is desired, it is perhaps best to obtain 25 dB per transducer. On the other hand, apodization cannot be used to implement the desired tap weights on both transducers without the added complexity of a multistrip coupler. The alternative method for implementing tap weights is withdrawal weighting, which has two limitations: (1) it can be used only for relatively narrowband responses with a large number of electrodes, and (2) the design algorithm is more difficult. A popular choice is to use withdrawal weighting on one transducer and apodization on the other; two apodized transducers with a multistrip coupler is also common.

Once the desired frequency response of each transducer is decided, the next step is to come up with an impulse response (for each transducer) of *finite time duration*. As we discussed in Section 1.4 a direct Fourier transform of the frequency response usually gives an impulse response of infinite time duration which has to be truncated appropriately so as to avoid excessive distortion. Various optimal schemes (such as the McClellan and Parks algorithm, Ref. 1.2) have been developed for determining the shortest impulse response that can meet a set of frequency specifications. Each transducer is then designed to produce the desired frequency response $\mu(f)$, as discussed in Section 4.3. This step is relatively straightforward for apodized transducers; the electrode lengths are proportional to the desired tap weights. For withdrawal-weighted transducers, an iterative scheme on a computer is required.

It is at this point that this chapter starts. We already have a preliminary design for each transducer; and the transfer function $\mu(f)$ and the admittance function $Y_a(f)$ of each can be determined using the methods presented in Sections 4.2 to 4.4. We now have to interface the transducers with external electrical circuitry, determine the insertion loss of the device as a function of frequency, and check how well the results satisfy the given specifications. This problem is discussed in Section 9.3. In Section 9.4 we discuss various second-order effects that can produce unwanted distortion and the techniques used to minimize them. We conclude this chapter with two illustrative examples of bandpass filter design in Section 9.5.

9.3. Analysis

As we mentioned earlier, the overall filter response is simply the product of the frequency responses of the input and output interdigital transducers (IDTs). The delay or propagation region between the transducers produces only an additional linear phase shift and makes no difference to the amplitude response. Of course, diffraction and other propagation losses do produce second-order effects (Section 9.4.2), which are neglected here. In this section we first consider the input (or transmitting) transducer and then the output (or receiving) transducer.

9.3.1. Transmitting IDT

Consider a transmitting IDT driven from a source with an impedance R_g. We will use the circuit model for IDT developed in Chapter 4 (Fig. 9.3). If voltage source were ideal with $R_g = 0$, the SAW power generated would be proportional to $V^2 G_a(f)$; since $G_a(f) \sim |\mu(f)|^2$, the frequency

9 BANDPASS FILTERS

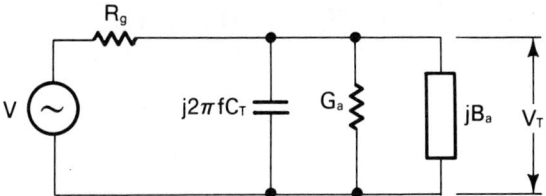

Figure 9.3 Equivalent circuit for an IDT driven with a source of impedance R_g.

response of the transducer is then determined by $\mu(f)$, as desired. However, an ideal voltage source has infinite available power, so that the insertion loss is infinite in this case. Real sources have finite available power; the available power, P_{av}, is the maximum power that can be drawn from the source, that is, the power delivered to a load under matched conditions.

$$P_{av} = \frac{V^2}{4R_g} \tag{9.1}$$

The actual voltage V_T appearing across the IDT is somewhat less than V because of the voltage drop in R_g.

$$\frac{V_T}{V} = \frac{1}{1 + Y(f)R_g} \tag{9.2a}$$

where

$$Y(f) = G_a(f) + jB_a(f) + j2\pi fC_T \tag{9.2b}$$

The SAW generated by the IDT has an amplitude proportional to V_T and not V. This introduces an additional frequency dependence, since Y is frequency dependent.

$$\phi^{\pm}(f) = \mu^{\pm}(f)H_m(f)V \tag{9.3a}$$

where

$$H_m(f) = \frac{1}{1 + Y(f)R_g} \tag{9.3b}$$

Thus the overall frequency response has two sources: $\mu(f)$ comes from the design of the IDT and is determined by the Fourier transform of the electrode voltages, while $H_m(f)$ is determined by external circuit conditions. Of course, since G_a and B_a depend on μ, the two are not really independent.

Let us now evaluate the insertion loss. This is defined as the power delivered to the SAW as a fraction of the available power. The power delivered to the forward-traveling SAW is $\frac{1}{2} V_T^2 G_a$, which is radiated equally in either direction. Since the receiver intercepts only one of the two beams, we take half of this power. Of course, this factor of $\frac{1}{2}$ would be absent for unidirectional transducers (UDTs). The insertion loss for bidirectional transducers then is

$$\text{IL (in dB)} = -10 \log_{10} \frac{\frac{1}{2} V_T^2 G_a}{V^2/4R_g}$$

$$= -10 \log_{10} \left[2 G_a R_g \, |H_m|^2\right] \tag{9.4a}$$

$$= -10 \log_{10} \frac{2 G_a R_g}{(1 + G_a R_g)^2 + \left[R_g(2\pi f C_T + B_a)\right]^2} \tag{9.4b}$$

Note that G_a and B_a are usually strongly frequency dependent compared to $2\pi f C_T$, so that if G_a (and B_a which is proportional to G_a) is much smaller than $2\pi f C_T$, we can neglect the frequency dependence of H_m over the passband. The frequency response of the IDT is then essentially that of G_a (that is, $|\mu|^2$). But if G_a is comparable to $2\pi f C_T$, H_m can contribute significantly, thus distorting the bandshape from that of $|\mu|^2$. The former is referred to as the weakly coupled case, and the later as the strongly coupled case. An important parameter for the IDT is its acoustic Q_a, defined as

$$Q_a = \left(\frac{2\pi f C_T}{G_a}\right)_{\text{at center frequency}} \tag{9.5}$$

A high Q_a signifies weak coupling; a low one signifies strong coupling.

Example 9.1
Calculate the Q_a for the solid- and split-electrode IDTs in Examples 4.15, 4.17, and 4.18, at the fundamental center frequency.

9 BANDPASS FILTERS

Solution
For the solid-electrode IDT,

$$\mu(f_o) = 0.8 \, jK^2 N \quad \text{(Example 4.5)}$$

$$G_a(f_o) = 1.28(K^2N)^2 Y_0 \tag{4.21}$$

$$C_T = NC_S W \quad \text{(Example 4.17)}$$

$$2\pi f_o C_T = K^2 N Y_0 \tag{3.22}$$

$$Q_a = \frac{0.78}{K^2 N} \tag{9.5}$$

The double-electrode IDT has the same G_a at the center frequency $(= f_o/2)$ as the single-electrode IDT (see Example 4.15), but it has a capacitance that is 1.4 times larger (see Example 4.18). Hence its Q_a is 1.4 times larger at the center frequency.

$$Q_a = \frac{1.1}{K^2 N}$$

In our examples, $K^2 = 0.046$ and $N = 10$.

$$Q_1 = \begin{matrix} 1.7 \\ 2.37 \end{matrix} \quad \begin{matrix} \text{for a solid-electrode IDT} \\ \text{for a split-electrode IDT} \end{matrix}$$

Note that these are the same values that we get if we calculate Q_a from Eq. 9.5 directly.

Example 9.2
Calculate the insertion loss of the solid-electrode unapodized IDT in Example 4.15 at its fundamental center frequency if it is driven from a 50-Ω generator. What is the shape of the insertion loss versus frequency?

Solution

$$R_g = 50 \, \Omega$$

$$G_a = 0.61 \text{ mmho} \quad \text{(Example 4.15)}$$

$$2\pi f_o C_T = 1.01 \text{ mmhos} \quad \text{(Example 4.17)}$$

$$B_a = 0 \quad \text{(at center frequency)}$$

$$\text{IL} = 12.4 \text{ dB} \quad \text{(From 9.4b)}$$

Of this 12.4 dB, 3 dB comes from the directionality of the IDT; half the power is radiated in the wrong direction. The other 9.4 dB represents mismatch loss; this power is reflected back to the generator due to the mismatched impedance. The IDT itself is lossless in our present model.

In this example both $G_a R_g$ and $R_g 2\pi f C_T$ are rather small, so that

$$\text{IL} \simeq -10 \log 2 G_a R_g$$

Hence the IL has roughly the shape of G_a, that is $(\sin^2 x)/x^2$ ($\sim \mu^2$).

The insertion loss equation has to be modified slightly if there is a loss conductance G_ℓ in parallel with G_a.

$$\text{IL} = -10 \log_{10} \frac{2 G_a R_g}{\left[1 + (G_a + G_\ell) R_b\right]^2 + \left[R_g(2\pi f C_T + B_a)\right]^2} \quad (9.6)$$

This loss conductance, G_ℓ, can arise because of ohmic losses in the electrodes or the generation of waves other than surface waves (Section 9.4.5). For apodized IDTs the G_{aA} representing power delivered to the zero-average components of the beam can be treated as a loss conductance (see Example 4.16).

Example 9.3
Calculate the insertion loss of the solid-electrode apodized IDT in Example 4.16 at its fundamental center frequency if it is driven from a 50-Ω generator.

Solution

$$R_g = 50 \text{ Ω}$$

$$G_a = G_{aU} = 0.19 \text{ mmho} \quad \text{(Example 4.16)}$$

$$G_\ell = G_{aA} = 0.06 \text{ mmho}$$

$$2\pi f C_T = 1.26 \text{ mmhos} \quad \text{(Example 4.19)}$$

9 BANDPASS FILTERS

$$B_a = 0$$

$$\text{IL} = 17.3 \text{ dB} \tag{9.6}$$

The apodization loss is defined as the ratio of the power delivered to the uniform component of the SAW to the total power in the SAW beam:

$$\text{apodization loss} = -10 \log_{10} \frac{G_{aU}}{G_{aU} + G_{aA}}$$

$$= 1.2 \text{ dB} \quad \text{in this case}$$

For the apodized IDT, the 17.3 dB of insertion loss is composed of 3 dB of bidirectional loss, 1.2 dB of apodization loss, and 13.1 dB of mismatch loss. The mismatch loss can be eliminated with proper matching circuits but not much can be done about the other two.

Conjugate matching completely eliminates the mismatch loss. In conjugate matching, a matching network is used to transform the IDT impedance to equal the generator impedance (at center frequency) and the generator impedance is made equal to the inverse of the radiation conductance. The insertion loss then becomes 3 dB plus any apodization loss. Triple-transit and higher-order reflections (Section 9.4.3) create large ripples in the passband of such a closely matched filter; at the ripple peaks the insertion loss can be less than 3 dB.

Example 9.4
(a) Calculate the inductance L and the generator impedance R_g needed to conjugate match the IDT in Example 9.2, using shunt and series tuning (Fig. 9.4 a and b).

Solution
Shunt tuning:

$$f = 100 \text{ MHz}$$

$$2\pi f C_T = 1.01 \text{ mmhos}$$

$$G_a = 0.61 \text{ mmho}$$

$$2\pi f L = \frac{1}{2\pi f C_T} \quad \text{at } f = 100 \text{ MHz}$$

$$L = 1.58 \ \mu\text{H}$$

$$R_g = 1.64 \ \text{K}\Omega$$

Series tuning:

$$2\pi f L = \frac{2\pi f C_T}{(G_a)^2 + (2\pi f C_T)^2}$$

$$L = 1.15 \ \mu\text{H}$$

$$R_g = \frac{G_a}{(G_a)^2 + (2\pi f C_T)^2}$$

$$= 438 \ \Omega$$

Note that series tuning requires a lower generator impedance than shunt tuning.

(b) Suppose that we have a 50-Ω generator. Calculate the L, C_1, and C_2 required to pi match the IDT in this example (Fig. 9.4c).

Solution
With a little bit of circuit analysis we can show that in order to get a conjugate match, we must have

$$\frac{C_2 - x_2 C_1}{G_a} = \frac{x_1 L G_a}{x_2}$$

and

$$G_a R_g = \frac{x_2}{x_1}$$

where $\quad C_2 = C_2' + C_T$

$$x_1 = 4\pi^2 f^2 L C_1 - 1$$

$$x_2 = 4\pi^2 f^2 L C_2 - 1$$

9 BANDPASS FILTERS

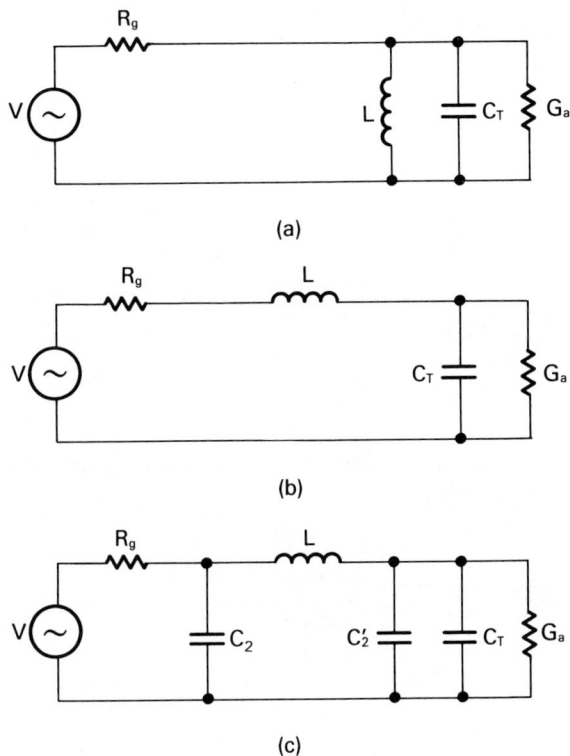

Figure 9.4 IDT transmitter conjugate at center frequency: (a) shunt tuning; (b) series tuning; (c) pi matching.

These conditions are satisfied if we choose

$$(2\pi f C_1 R_g)^2 + 1 = G_a R_g \left[1 + \left(\frac{2\pi f C_2}{G_a}\right)^2\right]$$

$$\frac{2\pi f L}{R_g} = \frac{2\pi f C_1 R_g + 2\pi f \dfrac{C_2}{G_a}}{(2\pi f C_1 R_g)^2 + 1}$$

$$f = 100 \text{ MHz}$$

$$2\pi f C_T = 1.01 \text{ mmhos}$$

$$G_a = 0.61 \text{ mmho}$$

$$R_g = 50 \text{ }\Omega$$

Let us choose $C_2' = 8.1$ pF, so that

$$\frac{2\pi f C_2'}{G_a} = 10$$

Then

$$2\pi f C_1 R_g = 1.44$$

$$C_1 = 45.91 \text{ pF}$$

$$L = 0.296 \text{ }\mu\text{H}$$

Under tuned conditions the insertion loss is obtained from Eq. 9.4b by leaving out the susceptive part.

$$\text{IL} = -10 \log_{10} \frac{2 G_a R_g}{(1 + G_a R_g)^2}$$

As we can see, this gives 3 dB if we put $R_g = \dfrac{1}{G_a}$. We mentioned in connection with Eq. 9.4 that the presence of $2\pi f C_T$ helps to keep the denominator nearly constant over the passband, by swamping out the rapid variations in G_a. But under tuned conditions this term is absent, so that the insertion loss versus frequency can deviate significantly from the $(\sin^2 x)/x^2$ shape, especially if we drive it mismatched with a high source impedance. Figure 9.5 shows the insertion loss versus frequency for the unapodized IDT in Example 4.15 for different values of $G_a R_g$ (at center

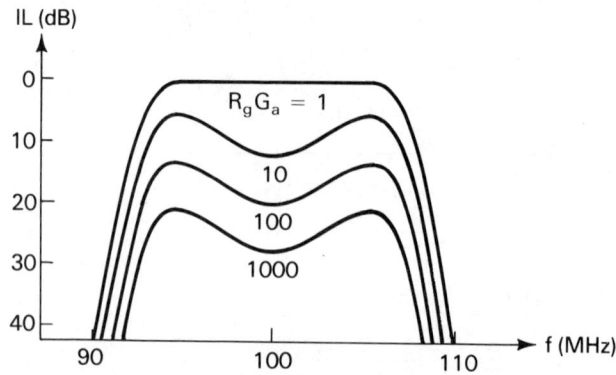

Figure 9.5 Passband shapes for tuned transducers with mismatch.

9 BANDPASS FILTERS

frequency). Note that in the matched case the passband is rather flat. Mechanical reflections from the electrodes are ignored here.

Example 9.5 Voltage Standing-Wave Ratio
Consider an IDT conjugate matched at its center frequency. What will be the VSWR in the transmission line connecting the device to the generator (a) at center frequency; (b) at a frequency where the insertion loss is 3 dB higher?

Solution
(a) Since the IDT is conjugate matched at the center frequency, there is no reflection from the device at this frequency. Hence the reflection coefficient $\rho = 0$ and the voltage standing-wave ratio

$$VSWR = \frac{1 + |\rho|}{1 - |\rho|} = 1$$

(b) At the 3-dB frequency, however, the IDT is not matched anymore. In fact, the extra 3 dB of insertion loss is because half the power is reflected back.

$$|\rho| = 0.707$$

$$VSWR = 5.8$$

If high VSWRs are unacceptable, losses in the form of electrical resistance should be added to the IDT.

Before we end this section on conjugate matching let us look at something that appears puzzling at first sight. At the center frequency, an IDT can be conjugate matched to give 3 dB of insertion loss, no matter what its Q_a is. In other words, we could use a very weakly piezoelectric solid (very small K^2) and use only a couple of electrodes, so that Q_a was very high, and still have only 3-dB of insertion loss. Why, then, do we want to use a strongly piezoelectric solid such as lithium niobate?

The answer is: We would like our IDT [as represented by $G_a(f)$] to determine the filter response and not the LC matching network. Note (Fig. 9.4) that the parallel LC_T circuit and the generator impedance R_g form a bandpass filter that is in cascade with the IDT. We would definitely like the IDT to be narrower in band than the external matching network, or else we really have an LC filter and the SAW device contributes nothing useful. The fractional bandwidth of the matching

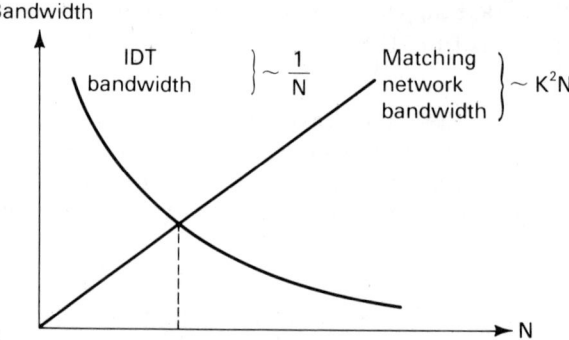

Figure 9.6 Bandwidths of IDT and conjugate matching network as a function of N.

network is $(2\pi f C_T R_g)^{-1}$. If $R_g = \dfrac{1}{G_a}$ (for conjugate match), this comes to $\dfrac{1}{Q_a} \sim K^2 N$. The fractional bandwidth of the IDT is $\dfrac{1}{N}$, so that we must have $\dfrac{1}{N} < (K^2)^{\frac{1}{2}}$ (Fig. 9.6); thus one can only build filters with fractional bandwidths less than $(K^2)^{\frac{1}{2}}$ for the substrate. This means about 20% bandwidth for lithium niobate but only about 2% for ST quartz. For narrowband filters, it is better to use quartz rather than lithium niobate, for a large number of electrodes on a strong piezoelectric causes unwanted second-order effects. Of course, ST quartz could also be used to build wideband filters if we did not insist on conjugate matching, but then we must put up with an insertion loss higher than 3 dB.

Because of the extra complication of adding an inductor (especially at UHF, where inductors often behave unpredictably), it is common to use magnitude matching rather than conjugate matching. This means that we use a generator impedance R_g or a load impedance R_L equal to the magnitude of transducer impedance:

$$R_{g,L}^{-1} = \left[(G_a)^2 + (2\pi f C_T)^2\right]^{\frac{1}{2}} \text{ at center frequency} \qquad (9.7)$$

From Eq. 9.4b, we then have for the insertion loss at the center frequency,

$$IL = -10 \log_{10} \frac{G_a R_g}{1 + G_a R_g} \qquad (9.8a)$$

$$= 10 \log_{10} \left(1 + \sqrt{1 + Q_a^2}\right) \qquad (9.8b)$$

9 BANDPASS FILTERS

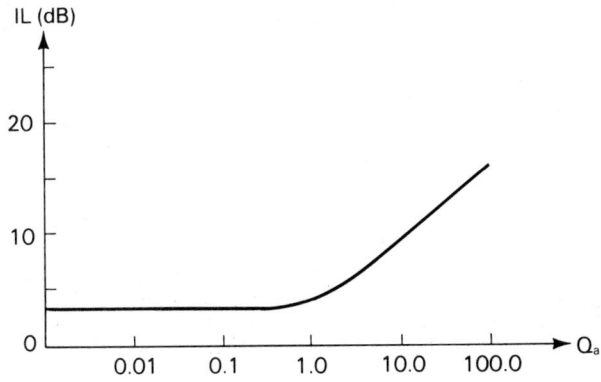

Figure 9.7 Magnitude-matched insertion loss versus Q_a.

where Q_a is defined by Eq. 9.5. The variation of the magnitude-matched insertion loss with Q_a is shown in Fig. 9.7. Note that if $Q_a \ll 1$, magnitude matching is just as good as conjugate matching, which is expected since if $G_a \gg \omega C_T$ we really do not need to tune out the capacitance.

9.3.2. Receiving IDT

The equivalent circuit for a receiving IDT with a load impedance R_L is shown in Fig. 9.8. The surface wave acts as a current generator $g_m \phi$ driving the IDT admittance in parallel with the load R_L. From reciprocity, we expect the insertion loss to be the same as that of a transmitting IDT with a source impedance $R_g = R_L$.

$$\text{IL} = -10 \log \frac{2 G_a R_L}{(1 + G_a R_L)^2 + \left[R_L (2\pi f C_T + B_a)\right]^2} \tag{9.9}$$

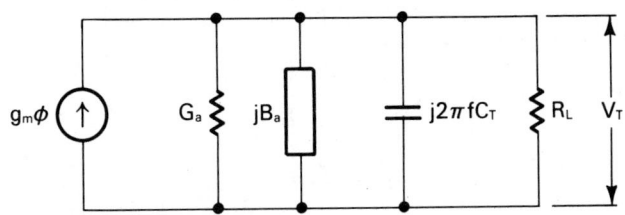

Figure 9.8 Equivalent circuit for an IDT receiver with a load impedance R_L.

Example 9.6

Consider the IDT in Example 4.21 used as a receiver with a load of 50 Ω. Show that the insertion loss is the same as when it is driven from a generator with an impedance of 50 Ω.

Solution

$$y_o = 0.22 \text{ mmho}$$

$$\frac{W}{\lambda} = 10$$

$$G_a = 0.61 \text{ mmho} \qquad \text{(Example 4.15)}$$

$$2\pi f C_T = 1.01 \text{ mmhos} \qquad \text{(Example 4.17)}$$

$$|\mu| = 0.37j \qquad \text{(Example 4.8)}$$

$$|g_m| = 1.64j \text{ mmhos} \tag{4.4}$$

Suppose that we have an incident SAW of amplitude 1 V. It carries a power P given by

$$P = 1.11 \; mW \tag{3.4, 3.7}$$

The voltage V_T induced between the transducer terminals (Fig. 9.9) is obtained from straightforward circuit analysis:

$$V_T = \frac{g_m \phi R_L}{(1 + G_a R_L) + j 2\pi f C_T R_L}$$

$$= 7.95 \times 10^{-1} \angle 87.2° \text{ volts}$$

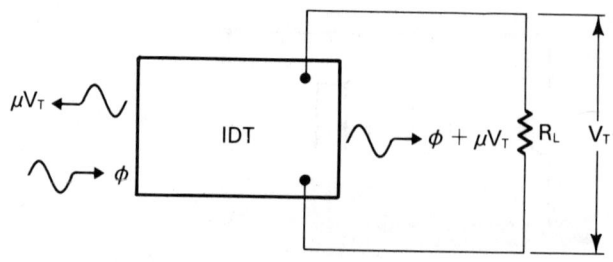

Figure 9.9 Acoustic-wave reflection and transmission at a loaded IDT receiver.

9 BANDPASS FILTERS

The power, P_L delivered to the load is

$$P_L = \frac{|V_T|^2}{2R_L}$$

$$= 0.063 \text{ mW}$$

$$\text{IL} = -10 \log \frac{P_L}{P} = 12.4 \text{ dB}$$

which is exactly the same as in Example 9.2.

We see here that 0.063 mW is delivered to the load out of a total of 1.11 mW. The logical question to ask is: What happens to the other 1.047 mW? The answer is that it is partially reflected and partially transmitted. Note that the receiving IDT has a voltage of V_T induced across its terminals. So it also functions as a transmitter generating a SAW of amplitude μV_T in either direction. This gives rise to the reflected wave while the transmitted wave is the sum of the original wave and the newly generated forward wave (Fig. 9.9). In Example 9.6 we can check that there is negligible reflection and most of the power is transmitted. As we will see, the reflection becomes significant under matched conditions (see Example 9.7).

Example 9.7
Suppose that we use the IDT in Example 9.6 as a receiver but under conjugate-matched conditions. Calculate how much power is delivered to the load and how much is reflected if a 1-V SAW is incident on it.

$$\phi = 1 \text{ V}$$

Solution
As we have seen in Example 9.6, a 1-V SAW carries 1.11 mW of power. Using the circuit model in Fig. 9.8 (leaving out the capacitor and setting $R_L = 1/G_a$),

$$V_T = \frac{g_m \phi}{2 G_a}$$

$$= 1.35j \text{ volts}$$

$$P_L = 0.55 \text{ mW}$$

The reflected wave has an amplitude of μV_T (Fig. 9.9):

$$\mu V_T = -0.5 \text{ V}$$

The transmitted wave has an amplitude of $\phi + \mu V_T$:

$$\phi + \mu V_T = 0.5 \text{ V}$$

Both the reflected and transmitted waves have half the amplitude of the incident wave, thus carry one-fourth the incident power. So in a conjugate-matched receiver, half the power goes to the load, one-fourth is reflected, and one-fourth is transmitted.

Example 9.8
Consider the apodized IDT in Example 9.3 used as a conjugate-matched receiver. Calculate the power delivered to the load and the reflected power.

Solution
Figure 9.10 shows the equivalent circuit for the conjugate-matched apodized receiver. It is the same as that for an unapodized IDT, except that the radiation conductance is split into G_{aU} and G_{aA}.

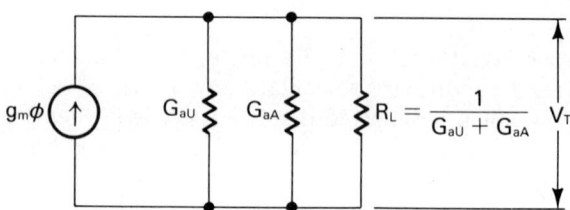

Figure 9.10 Equivalent circuit for a conjugate-matched apodized receiver.

If we assume that the incident SAW amplitude $\phi = 1$ V, we have

$$P = 5.6 \text{ mW} \hspace{4em} (3.40), (3.43)$$

Note that the power is five times larger than in the preceding examples because the width is 50 λ rather than 10 λ.

$$\mu = 0.093 j \hspace{2em} \text{(Example 4.14)}$$

$$g_m = 2.1 j \text{ mmhos}$$

9 BANDPASS FILTERS

$$G_{aU} = 0.19 \text{ mmho} \quad \text{(Example 4.16)}$$

$$G_{aA} = 0.06 \text{ mmho}$$

$$R_L = (G_{aU} + G_{aA})^{-1} = 4 \text{ k}\Omega$$

From straightforward circuit analysis,

$$V_T = 4.2j \text{ V}$$

$$P_L = 2.21 \text{ mW}$$

$$\text{IL} = -10 \log \frac{P_L}{P} = 4.1 \text{ dB}$$

Note that the insertion loss is higher than 3 dB because of apodization loss. The origin of this loss is not as obvious for a receiver as it is for a transmitter; but from reciprocity, if it is present in one, it has to be present in the other too.

$$\mu V_T = -0.39 \text{ V}$$

If we now use Eq. 3.4 to find the reflected power, we get 0.85 mW, which is only the power in the uniform component, since μV_T is the average amplitude of the nonuniform reflection. The total reflected power P_r is higher by the ratio $(G_{aU} + G_{aA})/G_{aU}$.

$$P_r = 1.12 \text{ mW}$$

By power conversation the transmitted power must be

$$P_t = 2.27 \text{ mW}$$

Note that in this case it is difficult to obtain the transmitted power directly. In principle, we have to add the uniform incident wave to the newly generated nonuniform wave and then calculate the power carried by the composite nonuniform wave.

Example 9.9
Consider the IDT in Example 9.6 used as a magnitude-matched receiver. Calculate the insertion loss and the reflected power.

Solution

$$R_L = 847 \ \Omega$$

Suppose that we have an incident SAW of amplitude 1 V carrying 1.11 mW as in Example 4.23. Following the same method gives us

$$V_T = 0.795 \angle 60° \text{ volts}$$

$$P_L = 0.37 \text{ mW}$$

$$IL = 4.7 \text{ dB}$$

Note that we get the same result if we use Eq. 4.40b with $Q_a = 1.7$ (Example 9.1).

$$\mu V_T = 0.29 \angle 150° \text{ volts}$$

$$\phi + \mu V_T = 0.76 \angle 11° \text{ volts}$$

Thus the reflected wave has an amplitude of 0.29 V and the transmitted wave has an amplitude of 0.76 V. The corresponding powers are obtained using Eq. 3.4.

$$P_r = 0.09 \text{ mW}$$

$$P_t = 0.64 \text{ mW}$$

Note that the insertion loss is only 1.7 dB higher than the conjugate matched case, but the reflected power is 5 dB lower (see Example 9.7). This is another reason for using magnitude matching. The reflected wave after another reflection at the transmitter interferes with the direct wave, giving what is known as the triple transit echo; this degrades filter performance by producing ripples in the passband. This is discussed in Section 9.4.3.

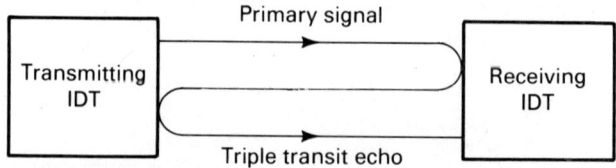

Figure 9.11 Triple transit echo in SAW filters.

9 BANDPASS FILTERS

9.4. Second-Order Effects

9.4.1. Internal Reflections

The frequency response of an IDT can be significantly distorted by multiple reflections of the SAW from the electrodes, which destroy its simple tapped delay line character. These reflections are often divided into two classes: regeneration and mechanical electrical loading (MEL). We have already discussed regeneration reflection in Section 9.3.2; an incident SAW induces a voltage at the IDT terminals, which in turn regenerates waves in both directions. The backward wave is called the regenerated wave and can be particularly large under matched conditions. We have also seen the distortion in frequency response resulting from generation; the insertion loss is not proportional to just $G_a(f)$ (Eq. 9.4), but is distorted somewhat. This distortion is particularly significant if the source resistance R_g is large compared to Y_a^{-1}.

If regeneration were the only source of reflection, we would expect no reflection or distortion, as R_g is reduced to zero. This is indeed the case for split-electrode transducers. The center-to-center distance between successive electrodes is a quarter of a wavelength at center frequency, so that the MEL reflections (arising from the discontinuity presented by the electrodes) from successive electrodes cancel. However, for solid-electrode transducers (often used to save on photolithographic resolution) the MEL reflections add up and can cause significant distortion even if R_g is small. This distortion shows up as a distortion in $G_a(f)$ itself which is no longer proportional to $|\mu(f)|^2$; it can be calculated numerically from a transmission-line model using different characteristic impedances (Z and $Z + \Delta Z$) for electrodes and free-surface regions, as discussed in Section 4.4.

9.4.2. Diffraction and Propagation Losses

The propagation of the surface wave from the input to the output transducer should ideally produce only a time delay with no distortion of the signal. In the frequency domain this means a linear phase response and a constant-amplitude response of 1. In practice, however, there is some loss of signal because diffraction causes the beam to widen and some of it misses the output transducer. This can lead to a slight increase in insertion loss.

A more serious problem, however, is that the amplitude and phase of the received signal depends on the propagation distance, so that it is different from different electrodes. This causes errors in the impulse response which show up as a reduced out-of-band rejection in the frequency response since a high rejection level requires precision in tap-weight implementation. Diffraction errors can, however, be compensated for in the design of transducers (Ref. 9.1). The basic idea is as follows. Consider one electrode of the transmitter and one electrode of the receiver.

Our usual assumption is that the received signal has a phase delay with respect to the input signal of kL, where $k = 2\pi/\lambda$ and L is the center-to-center distance between the electrodes. This is not strictly true. If we take any point X on the transmitter and only point Y on the receiver, the phase delay of the signal from X to Y is given by kR. The total output signal is obtained by summing over all points X and Y (Fig. 9.12):

$$\frac{I}{V} \propto \sum_{X,Y} \frac{e^{-jkR}}{\sqrt{R}} \tag{9.10}$$

Since R is in general not equal to L (unless X and Y "face" each other), it is clear that (I/V) is not proportional to e^{-jkL}; in fact, since $R \geq L$, the actual phase delay is always greater than kL. However, the absolute phase delay makes no difference to the frequency response; what does make a difference is that this *phase delay is different for different pairs of electrodes.* This is particularly true of apodized transducers, where each electrode has a different length and looks at different portions of the beam. The position of the electrodes can be adjusted to compensate for this distortion. For practical substrates the anisotropy has to be considered in evaluating Eq. 9.10; k is then a function of θ.

We have mentioned before (Section 9.3) that only certain selected cuts and orientation of different materials are acceptable as substrate materials. The reason is that in directions other than these pure-mode directions, diffraction effects are particularly severe; we also have beam steering, so that the direction of energy flow is not normal to the phase front. On such cuts and orientation, diffraction is not a second-order effect; it is a serious first-order effect.

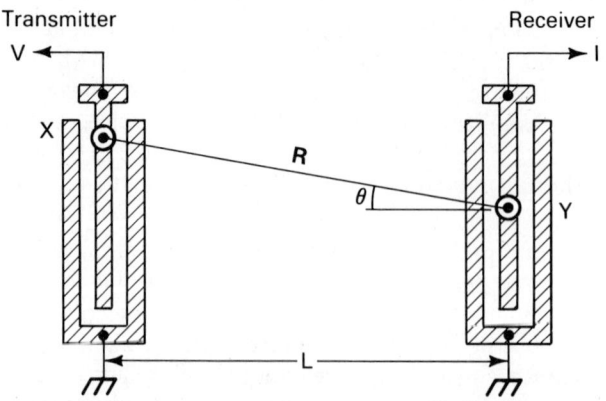

Figure 9.12 Diffraction effects of SAW filters.

9 BANDPASS FILTERS

9.4.3. Triple Transit Echo

We have seen in Section 9.3.2 that a significant fraction of the incident power is reflected from the receiving IDT when it is closely matched; with conjugate matching there is 25% reflection. If both transmitting and receiving IDTs are closely matched, this reflected signal can be reflected again and reach the receiving IDT after a delay equal to thrice the delay of the original signal (Fig. 9.11). This is known as the *triple transit echo* (TTE). Its magnitude relative to the original signal depends on the matching conditions at the input and output IDTs; if both are conjugate matched, the TTE is a maximum (12 dB below the original). This can give rise to significant ripples in the amplitude and phase response. To reduce such ripples, SAW devices are often operated in less than ideal conditions of matching, leading to an insertion loss greater than the minimum of 6 dB.

9.4.4. End Effects

End effects refer to the change in tap weight value due to fringing electrostatic fields not accounted for in a first-order analysis. For example, the field distribution near the last few electrodes in an array is different from that in the middle of the array, leading to a modification of the tap amplitude and phase. As discussed in Section 9.4, these effects have to be properly accounted for; this is especially true in the design of withdrawal-weighted transducers, where the periodicity of the array is repeatedly broken (Refs. 9.2 and 9.3).

A different effect, called the transverse end effect, is encountered in apodized transducers (Ref. 9.4). The strength of a tap is usually assumed to be proportional to the length of the electrode. However, electrostatic fields fringe out somewhat beyond the physical length of the electrode; if all electrodes were of the same length, this would affect each one equally. But in apodized transducers the effect is larger for shorter electrodes, causing a distortion in the impulse response. This factor often limits the "dynamic range" of tap weights that can be implemented by apodization.

9.4.5. Spurious Losses

Spurious losses are modeled by a loss conductance in parallel with the radiation conductance. There are two common sources of loss. The first is the ohmic loss due to current flow through the electrodes; this loss is fairly independent of frequency and can be reduced by using thicker films for electrodes. The other source of loss is the generation of bulk waves; this loss is frequency dependent and usually occurs on the high-frequency side of the passband since bulk waves are faster than surface waves (Ref. 9.5). These losses often do not distort the overall device response because the bulk waves generated by the input transducer travel away from the surface and are not seen by the output transducer. To ensure this, the back surface of the substrate is roughened, so that there is no coherent reflection of bulk waves.

9.5. Illustrative Examples of Filter Design

9.5.1. Wideband Filter on Lithium Niobate[*]

Specifications.

Center frequency	70.0 MHz
BW 1 dB minimum	10.0 MHz
BW 3 dB	
Nominal	11.4 MHz
Maximum	12.0 MHz
BW 50 dB	
Nominal	14.4 MHz
Maximum	15.0 MHz
Ultimate rejection	55 dB minimum
Amplitude ripple	0.25 dB
Phase ripple	2.0 deg
Insertion loss	25 dB maximum
System impedance	50 Ω

Configuration. The substrate material chosen for the filter is 128°-rotated Y-cut $LiNbO_3$. The reasons for choosing this material are (1) a high piezoelectric coupling constant, (2) low bulk-mode coupling, (3) a parabolic velocity surface, and (4) a relatively low-temperature coefficient (72 ppm/ °C). Because of high piezoelectric coupling coefficient, the discontinuity between metallized and free surface regions of the filter structure produces a large reflection only 38 dB below the main response; special efforts have to be made to reduce such spurious response. Also, the physical size of the structure must be made as small as possible to maximize the number of devices per wafer and minimize the effects of photolithographic defect densities.

Figure 9.13 shows the structure chosen for the design. The filter consists of two apodized transducers coupled with a multistrip coupler (MSC). The metallic strips for the transducers and the MSC all have the same linewidth and periodicity. Reflection suppression is achieved at the MSC-free surface boundaries by using reflection cancellation strips. Reflections at the ends of the transducers are reduced by tilting the ground busbars. At least a 40-wavelength separation between the transducer and the edge of the crystal is chosen to allow sufficient distance for proper acoustic damping. The MSC structure is a pair of 3-dB

---

[*]Courtesy of Phonon Corporation.

9 BANDPASS FILTERS

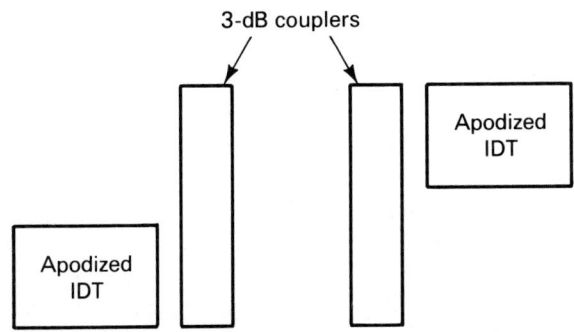

Figure 9.13 Structure of MSC-coupled bandpass filter.

couplers separated by a grounded electrostatic shield. The 3-dB couplers are identical to those used by Browning and Marshall (Ref. 9.6) incorporating a fanned "nose." We can split the filter analysis into two parts, each consisting of a transducer and a 3-dB coupler.

Design.
  Ideal Impulse Response Design. The ideal filter response is designed using the Remez exchange algorithm to obtain the shortest possible impulse response realization for the filter. The ideal design was chosen to have 0.07 dB of passband ripple and 65 dB of rejection. These figures are chosen to be somewhat better than the specifications to allow for realization errors.

The ideal impulse response is then deconvolved into two impulse responses (see Fig. 9.14a and b), using an automatic zero search algorithm developed at Phonon. This algorithm has been successfully used to deconvolve a 781-wavelength impulse response with better than 0.01-dB accuracy. The deconvolution results in the shortest possible filter length. This process produces a minimum phase–maximal phase transducer pair. It has been found that this method has better sidelobe sensitivity than a deconvolution into symmetric responses.

  Transducer Design. The transducers are apodized so as to implement the desired impulse responses as accurately as possible. The transducers and the MSC are chosen to have the same strip period of 16 μm, corresponding to a sampling frequency of 242.62 MHz. This choice of sampling frequency places the filter passband response below the MSC stopped and above the transducer split-finger stopband. The apodizations are shown in Fig. 9.14c. The transducer acoustic aperture was chosen to obtain the required insertion loss.

  Analysis. The analysis is crucial in the design of SAW filters. It should be as complete and efficient as possible. Phonon's analysis software includes the effects of attenuation, mass loading, strip resistance,

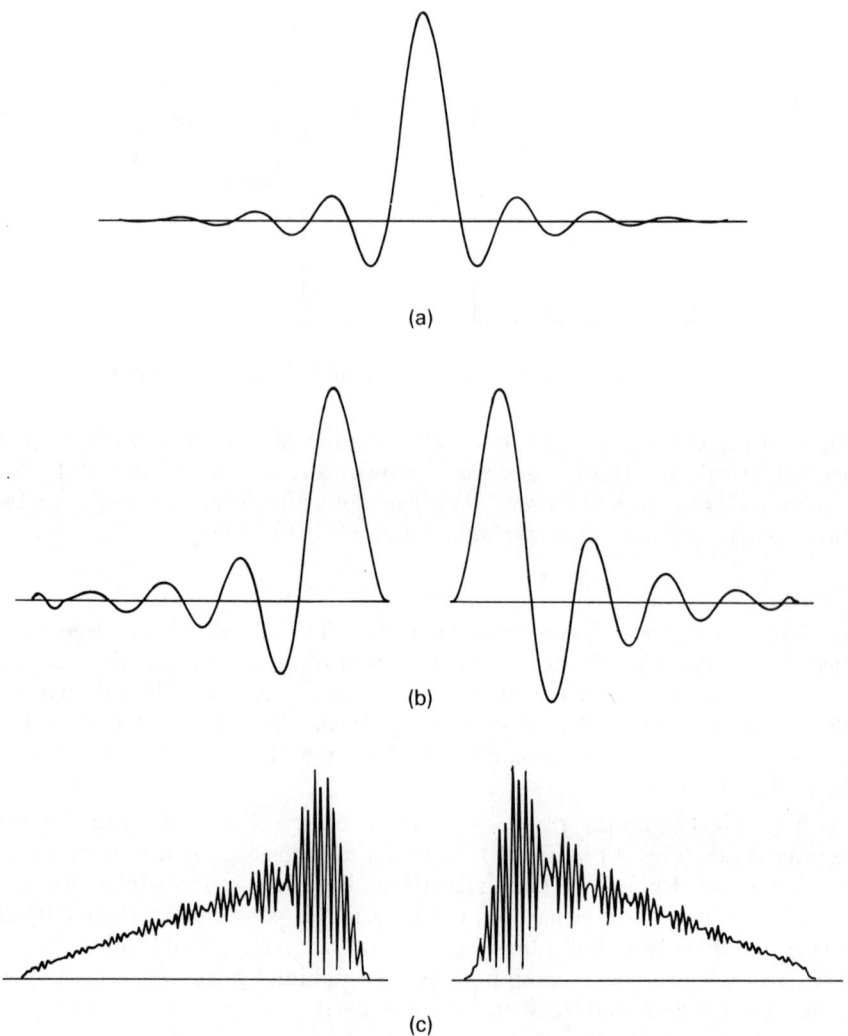

Figure 9.14 (a) Ideal impulse response of filter; (b) maximal phase and minimum phase impulse responses deconvolved from (a); (c) apodization of impulse responses in (b).

diffraction, topographic effects, and electrical matching networks in the computation of transducer impedance and the frequency-domain transfer function. In this case the analysis is split into two parts, each consisting of a transducer and a 3-dB coupler. Using this approach, the transducers can be individually corrected.

Transducer corrections can be performed either in the frequency domain or in the time domain; however, frequency sidelobe errors are often

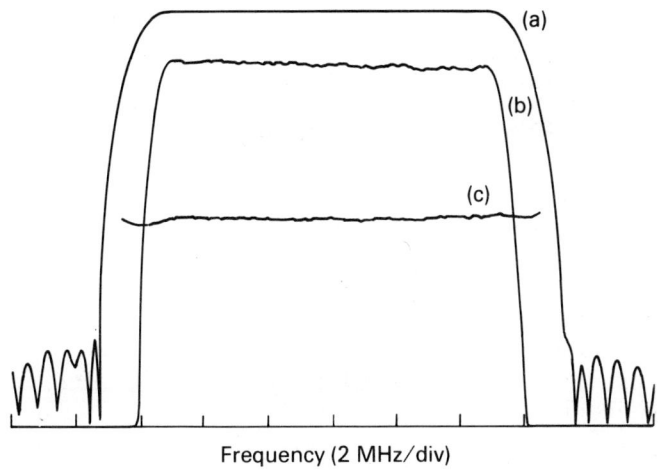

Figure 9.15 Predicted filter response: (a) insertion loss (10 dB/div); (b) insertion loss (1 dB/div); (c) phase (10 deg/div). References: loss, 24.27 dB; phase, -147.65°; delay, 1,296 μs; center frequency, 70 MHz.

impossible to correct in the frequency domain. The first correction for the two transducers was performed in the frequency domain to compensate for electrical loading, MSC coupling, and the frequency-dependent electroacoustic coupling characteristic. Subsequent corrections were performed in the time domain for diffraction compensation.

Measured Performance. Figure 9.15 shows the predicted response for the final design; Fig. 9.16 shows the measured frequency response.

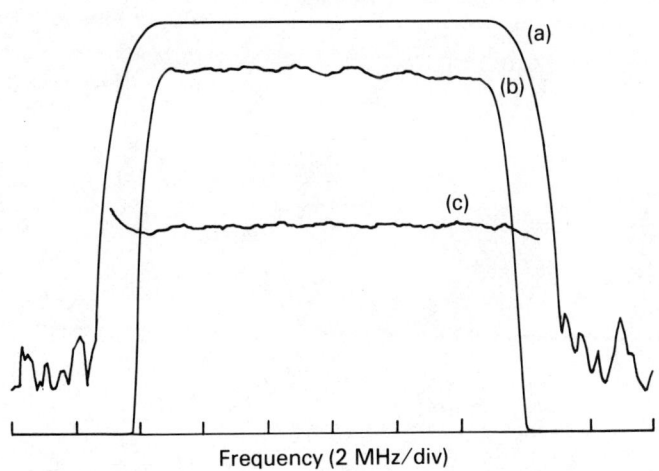

Figure 9.16 Measured filter response: (a) insertion loss (10 dB/div); (b) insertion loss (1 dB/div); (c) phase (10 deg/div). References: loss, 24.08 dB; phase, 65.65°; delay; 1.31 μs; center frequency, 70 MHz.

9.5.2. Narrowband Filter on ST-X Quartz[*]

Specifications. IF bandpass filter in a radio terminal set:

Center frequency	71.712 MHz
BW 1 dB minimum	1.9 MHz
BW 3 dB	
Minimum	2.9 MHz
Maximum	3.2 MHZ
BW 20 dB	
Minimum	4.43 MHz
Maximum	4.91 MHz
BW 35 dB	
Minimum	4.99 MHz
Maximum	5.54 MHz

Configuration. The design (Fig. 9.17) consists of an apodized and a uniform transducer with split fingers operating at the fundamental frequency. The substrate is 0.030-in.-thick ST-X quartz. Bulk-wave suppression is achieved by using two parallel acoustic channels with a

Figure 9.17 Enlarged picture of actual device.

---

[*]Courtesy of Andersen Laboratories.

9 BANDPASS FILTERS

phase difference of 180°; damping material is applied across one channel to suppress the surface wave, allowing the bulk wave to propagate and cancel at the output transducer. The bottom surface was not roughened. No careful data were taken, but the response at 1.6 to 1.8 times the center frequency was suppressed about 5 dB using this cancellation scheme. Low-pass tuning was used to ensure a broadband rejection of 35 dB.

Design. The ideal impulse response for the apodized transducer was synthesized using the McClellan et al. program (Ref. 1.2) for FIR linear phase filters with minimum weighted Chebyshev error in approximating an ideal bandpass function. Passband amplitude tilt and the passband weighting of the uniform transducer were precompensated by modifying the ideal passband function specified in the EFF subroutine. Sampling the impulse response at four times the center frequency allows the asymmetric bandshape to be achieved with a real impulse response. The resulting apodization pattern has a quadrature component introduced by the asymmetric passband.

Analysis. The analysis program computes the transducer capacitance and the channelized conductance based on the familiar superposition of solutions for the electrostatic field for one finger at unit potential in an infinite grounded array. The transfer function between transducers includes attenuation and diffraction, treating the short transducer as a line source. The transducer susceptance is calculated via the Hilbert transform of the conductance. The *ABCD* matrices for tuning elements, parasitics, transducer admittances, and the acoustic transmission line are cascaded to predict the filter insertion loss at the electrical port, as

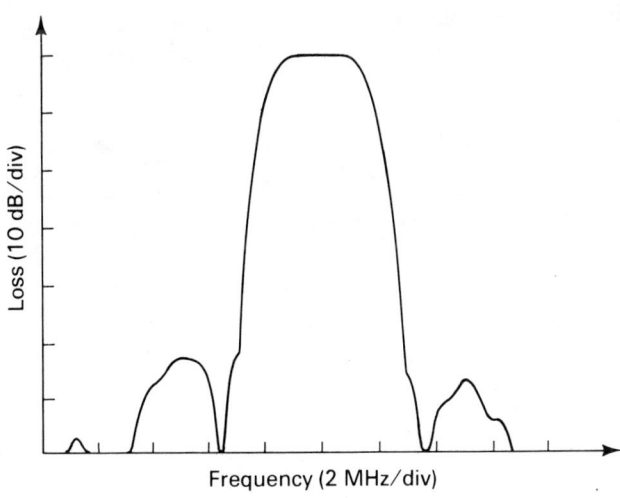

Figure 9.18 Predicted filter response.

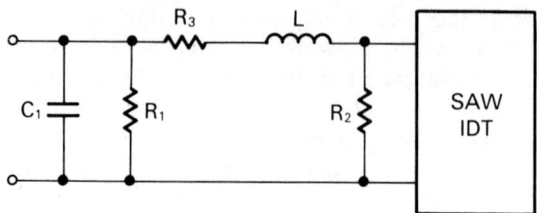

Figure 9.19 Tuning scheme.

shown in Fig. 9.18. The tuning network is shown in Fig. 9.19 with the theoretical and actual values for the uniform transducer.

Measured Performance. Device performance was measured using standard acceptance test procedures. Figure 9.20 shows the measured filter response. The measured insertion loss was 33.2 dB, compared with the predicted value of 29.8 dB. To obtain the correct center frequency, the aluminum film thickness was reduced from 2500 Å to 1800 Å.

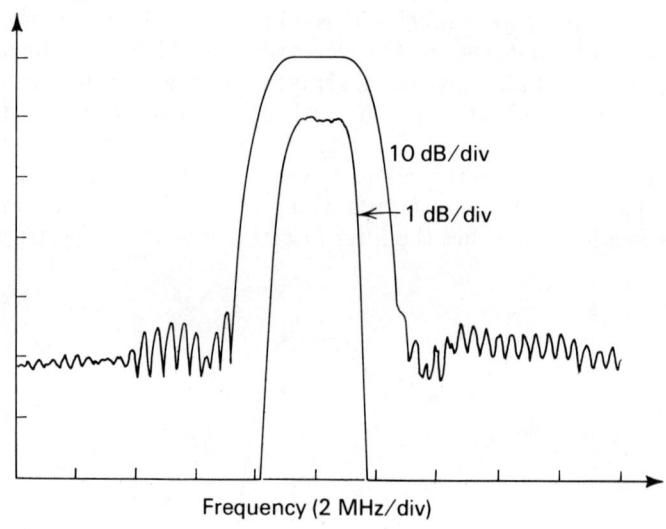

Figure 9.20 Measured filter response.

10
RESONATORS

We have seen in Chapter 6 that periodic arrays of grooves or electrodes can be used to construct efficient narrowband reflectors for surface waves. Two such reflectors at two ends of the SAW propagation path form a resonant cavity that can be used to make a resonator. In one-port resonators, a single IDT is placed between the reflectors; the input admittance of the IDT appears like a series resonant circuit in parallel (Fig. 10.1), with the transducer capacitance. In two-port resonators, two IDTs are placed in the cavity and the transfer function between them shows resonant behavior (Fig. 10.2). The two-port resonator is inherently

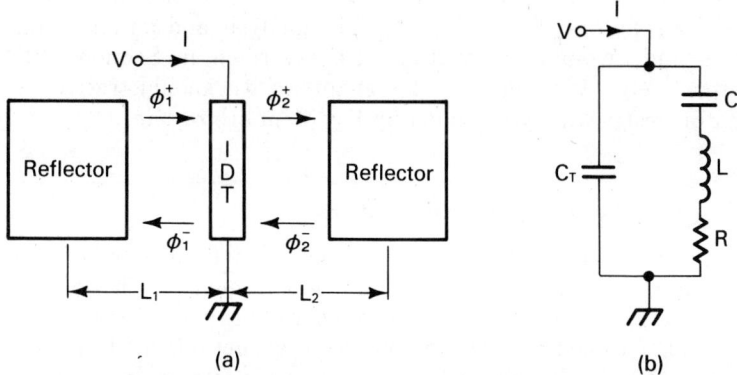

Figure 10.1 One-port resonator: (a) configuration; (b) equivalent circuit.

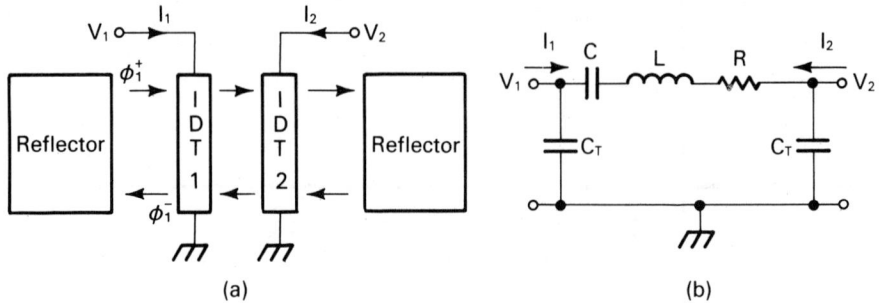

Figure 10.2 Two-port resonator: (a) configuration; (b) equivalent circuit.

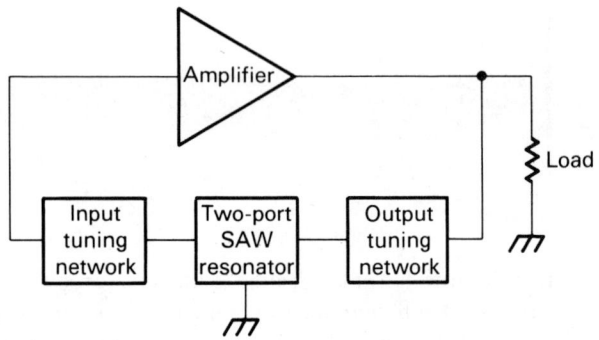

Figure 10.3 Block diagram of a SAW oscillator.

easier to use because the series resonant arm is not shunted by the transducer capacitance. A two-port resonator can be used to make an oscillator by including it in the feedback loop of an external amplifier (Fig. 10.3).

In Section 10.1 we describe the kind of specifications that can be met with SAW resonators and oscillators. The analysis and design of one-port resonators and two-port resonators are described in Sections 10.2 and 10.3, respectively. We conclude this chapter with an illustrative example of resonator design kindly provided by RF Monolithics, Inc.

10.1. Capabilities of SAW Resonators

Resonators are devices with a very narrow band frequency response and consequently a long impulse response. At low frequencies the long impulse response is obtained from the transfer of energy back and forth between an inductor and a capacitor. At microwave frequencies (~10 GHz) the long impulse response is obtained by generating an electromagnetic wave in a

10 RESONATORS

cavity several wavelengths long, where it bounces back and forth between the walls. The Q of the resonator is proportional to the number of wavelengths in the cavity and the number of times a wave bounces back and forth before its energy is dissipated. SAW two-port resonators are widely used for resonators and oscillators in the frequency range 100 to 1000 MHz, where conventional LC circuits start to fail due to parasitics, but the wavelength is not short enough for microwave cavities to be convenient. Acoustic waves, because of their slow velocity, have a wavelength 10^5 times shorter than an electromagnetic wave with the same frequency. Resonators using bulk acoustic waves have long been used at frequencies of 100 kHz to 50 MHz. But they are inconvenient to use at higher frequencies. The length of a SAW resonator is typically about 1 cm at 100 MHz and 1 mm at 1000 MHz. The length is basically determined by the lengths of the two reflectors that form the cavity. SAW reflectors are distributed reflectors about 100 to 200 wavelengths long, so that the length of a resonator is about 400 wavelengths. This factor limits the utility of SAW resonators at low frequencies, while at high frequencies propagation losses limit the available Q. Figure 10.4 shows some standard resonator specifications from RF Monolithics over the frequency range 200 to 1000 MHz. SAW resonators are usually fabricated on ST-cut quartz which has a high-temperature stability relative to other substrates. The frequency deviation is about 80 ppm over a $\pm 50°C$ temperature range.

10.2. One-Port Resonator

Consider the IDT in Fig. 10.1a positioned between two reflectors. We would like to eliminate the SAW variables $(\phi_1^\pm, \phi_2^\pm)$ to obtain an expression for the IDT admittance $(= I/V)$ from which we can find the equivalent circuit of Fig. 10.1b. The transducer capacitance C_T is of course always present; we will consider only the acoustic admittance which is in shunt with C_T.

The appropriate equations for the IDT are (neglecting any reflections due to the IDT)

$$\phi_2^+ = \phi_1^+ + \mu V \tag{10.1a}$$

$$\phi_1^- = \phi_2^- + \mu V \tag{10.1b}$$

$$I = -g_m \phi_1^+ - g_m \phi_2^- + GV \tag{10.1c}$$

where μ and g_m are the transmitter and receiver response functions and G is the radiation conductance of the IDT *if no reflectors were present*. In that case $\phi_1^+ = \phi_2^- = 0$ and we get the usual result:

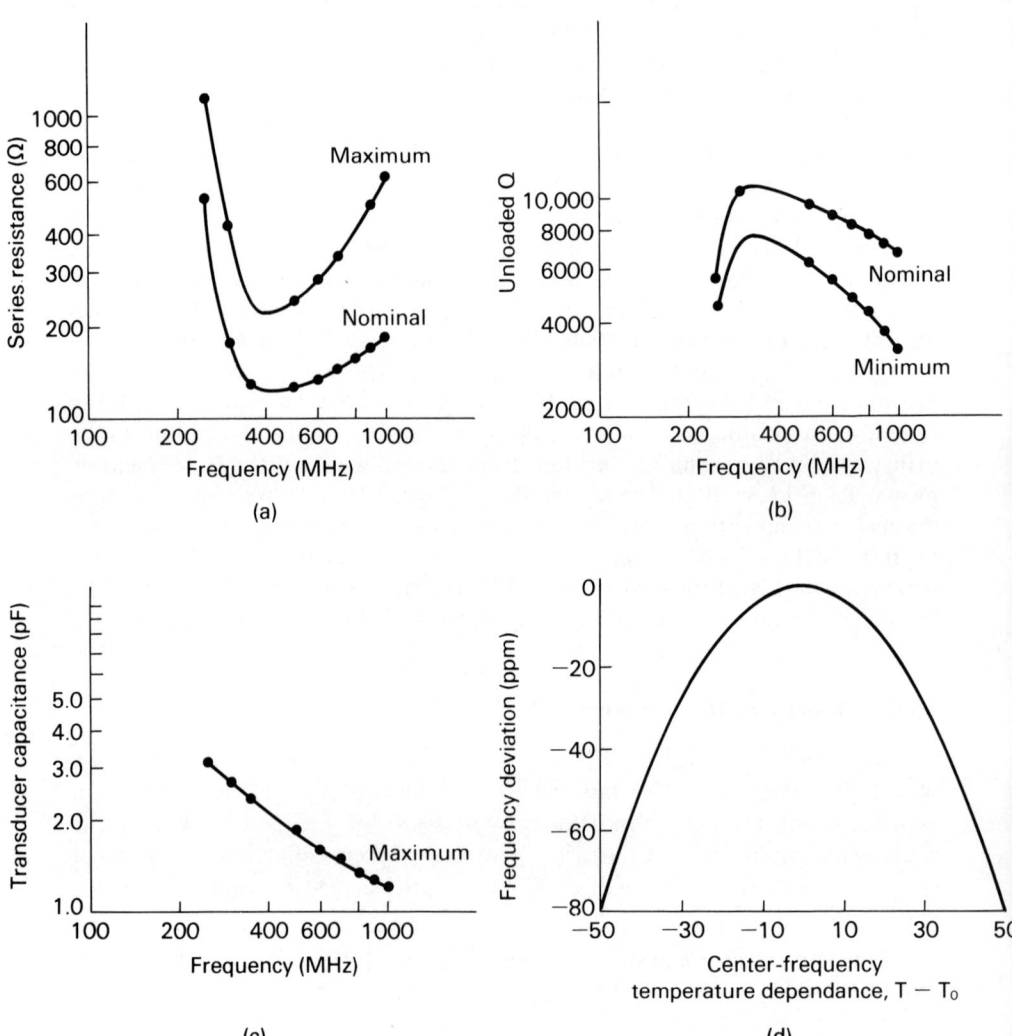

Figure 10.4 RF Monolithics' standard resonator specifications: (a) nominal and maximum values of series resistance; (b) nominal and minimum values of unloaded Q; (c) nominal terminal capacitance; (d) frequency deviation of a SAW resonator fabricated on ST-cut quartz as a function of temperature. T_0 is the turnover temperature.

10 RESONATORS

$$\phi_2^+ = \phi_1^- = \mu V$$

$$I = GV$$

However, because of the reflectors we have reflected waves from either side, so that

$$\phi_1^+ = \Gamma_1 \phi_1^- \tag{10.2a}$$

$$\phi_2^- = \Gamma_2 \phi_2^+ \tag{10.2b}$$

where Γ_1 and Γ_2 are the reflection coefficients of reflectors 1 and 2 respectively, *referenced to the center of the IDT,* where μ and g_m are also referenced. Over a range of frequencies around the center frequency both Γ_1 and Γ_2 have a magnitude close to 1 and the phase varies approximately linearly with frequency (Fig. 6.5), the phase slope depending on the lengths L_1 and L_2 from the reference plane to the effective center of reflection inside the reflector (Fig. 6.4).

$$\Gamma_1 = \Gamma e^{-j\theta_1} \tag{10.3a}$$

$$\Gamma_2 = \Gamma e^{-j\theta_2} \tag{10.3b}$$

where $\Gamma \simeq 1$ and where θ_1 and θ_2 vary linearly with frequency

$$\theta_1 = \frac{4\pi L_1}{\lambda} \pm \frac{\pi}{2} \tag{10.4a}$$

$$\theta_2 = \frac{4\pi L_2}{\lambda} \pm \frac{\pi}{2} \tag{10.4b}$$

The $\pm \frac{\pi}{2}$ comes from the fact that all lossless reflectors (such as inductors and capacitors in a transmission line) have a reflection coefficient whose phase is either $+\frac{\pi}{2}$ or $-\frac{\pi}{2}$, when referenced to the center of the reflector. Since reflectors are usually $\frac{\lambda}{4}$ wide, this means that the phase of the reflection coefficient is either 0 or π when referenced to the reflector edge (see Example 6.3). L_1 and L_2 are rounded off to the nearest wavelength so that $L_{1,2}/\lambda$ is an integer. Otherwise, the extra phase factor will not be

$L\pi$; it will have a value such that $\theta_{1,2}$ comes out as $\pm\dfrac{\pi}{2}$ at center frequency.

Let us now get back to the original problem of eliminating the SAW variables $\phi_{1,2}^{\pm}$ from Eq. 10.1 so as to get an expression for the IDT admittance. Using Eqs. 10.2a and 10.2b in Eqs. 10.1a and 10.1b, we get

$$\phi_1^+ = \mu V \frac{\Gamma_1(1+\Gamma_2)}{1-\Gamma_1\Gamma_2} \tag{10.5a}$$

$$\phi_2^- = \mu V \frac{\Gamma_2(1+\Gamma_1)}{1-\Gamma_1\Gamma_2} \tag{10.5b}$$

Using Eqs. 10.5a and 10.5b in Eq. 10.1c, we have

$$I = GV - g_m \mu V \frac{\Gamma_1(1+\Gamma_2) + \Gamma_2(1+\Gamma_1)}{1-\Gamma_1\Gamma_2} \tag{10.6}$$

Using the relation $G = -g_m\mu$, we can combine the two terms in Eq. 10.6 to give

$$Y = \frac{I}{V} = G \frac{(1+\Gamma_1)(1+\Gamma_2)}{1-\Gamma_1\Gamma_2} \tag{10.7a}$$

where Y is the admittance of the IDT. Note that G is the radiation conductance we would see if there were no reflectors. This is easily checked by setting $\Gamma_1 = \Gamma_2 = 0$ in Eq. 10.7a. Y represents the admittance of the arm in parallel with the transducer capacitance (Fig. 10.1b). It is a little more convenient to work with the impedance Z of this arm because of its series resonant nature.

$$Z = \frac{1}{G} \frac{1-\Gamma_1\Gamma_2}{(1+\Gamma_1)(1+\Gamma_2)} \tag{10.7b}$$

From Eq. 10.7b it is clear that the condition of series resonance is given by $\Gamma_1\Gamma_2 \simeq 1$, so that from Eqs. 10.3a and 10.3b we must have

$$\theta_1 + \theta_2 = 2n\pi$$

where n is an integer. Using Eqs. 10.4a and 10.4b, we get

$$\frac{2L_1 + 2L_2 \pm \lambda}{2} = n\lambda \qquad (10.8)$$

This says that the total effective length of a round trip in the cavity must be an integer number of wavelengths. The $\pm \frac{\lambda}{2}$ comes from the extra phase shift of $\pm \frac{\pi}{2}$ associated with each reflection.

To minimize the series resistance at resonance we should also maximize the denominator so that $\Gamma_1 = \Gamma_2 \simeq 1$.

$$\theta_1 = 2m_1 \pi$$

$$\theta_2 = 2m_2 \pi$$

where m_1 and m_2 are integers. Using Eqs. 10.4a and 10.4b yields

$$2L_1 \pm \frac{\lambda}{4} = m_1 \lambda \qquad (10.9a)$$

$$2L_2 \pm \frac{\lambda}{4} = m_2 \lambda \qquad (10.9b)$$

Note that any two of the three conditions 10.8, 10.9a, and 10.9b imply the other. The conditions expressed in Eqs. 10.9a and 10.9b ensure that the transducer is located on a maximum of the standing-wave pattern set up in the cavity. At center-frequency conditions 10.8 and 10.9 are satisfied, so that the impedance Z (Eq. 10.7b) is purely resistive and equal to the series resistance R in the equivalent circuit (Fig. 10.1b).

$$R = \frac{1}{G} \frac{1 - \Gamma}{1 + \Gamma} \qquad (10.10)$$

Γ being the magnitude of the reflection coefficient of either reflector (Eq. 10.3).

The variation of the impedance Z with frequency is really very different from that of an LCR circuit as we can easily see from Eq. 10.7b.

$$Z = \frac{1}{G} \frac{1 - \Gamma^2 e^{-j(\theta_1 + \theta_2)}}{(1 + \Gamma e^{-j\theta_1})(1 + \Gamma e^{-j\theta_2})} \qquad (10.11a)$$

where θ_1 and θ_2 depend on frequency through Eq. 10.4. By contrast, the impedance Z' of an LCR circuit is given by

$$Z' = R + j\left(2\pi f L - \frac{1}{2\pi f C}\right) \qquad (10.11\text{b})$$

However, we can choose L and C such that Z and Z' agree over a small range of frequencies Δf around the center frequency. For small Δf we can write

$$Z = \frac{1}{G} \frac{(1 - \Gamma^2) + j\Gamma^2 \, 4\pi \Delta f (L_1 + L_2)/v_o}{(1 + \Gamma)^2} \qquad (10.12\text{a})$$

$$Z' = R + j4\pi \, \Delta f \, L \qquad (10.12\text{b})$$

where we have used $\lambda f = v_o$. Comparing Eqs. 10.12a and 10.12b, we have

$$R = \frac{1}{G} \frac{1-\Gamma}{1+\Gamma} \quad \text{(as before)} \qquad (10.13\text{a})$$

$$L = \frac{1}{4Gf_o} \frac{L_1 + L_2}{\lambda} \qquad (10.13\text{b})$$

since $\Gamma \simeq 1$. Also,

$$C = \frac{1}{4\pi^2 f_o^2 L} \qquad (10.13\text{c})$$

Note that the LCR equivalent is valid only around the center frequency but the impedance Z can be calculated numerically over a wide band of frequencies directly from Eq. 10.7b using the proper $G(f)$, $\Gamma_1(f)$, and $\Gamma_2(f)$. The Q (unloaded) is given by

$$Q = \frac{2\pi f_o L}{R} = \frac{\pi}{1-\Gamma} \frac{L_1 + L_2}{\lambda} \qquad (10.14)$$

Example 10.1
Consider a cavity formed by grooved array reflectors on ST quartz (Fig. 10.5). With an IDT in the center, check to see if conditions 10.8 and 10.9 are satisfied. What is the center frequency of the resonator? Calculate the series resistance R and the unloaded Q assuming that there are 100 strips in each reflector array and that the IDT has four pairs of solid electrodes and is 0.16 mm wide.

$$\lambda = 2 \times 4 \, \mu\text{m} = 8 \, \mu\text{m}$$

10 RESONATORS

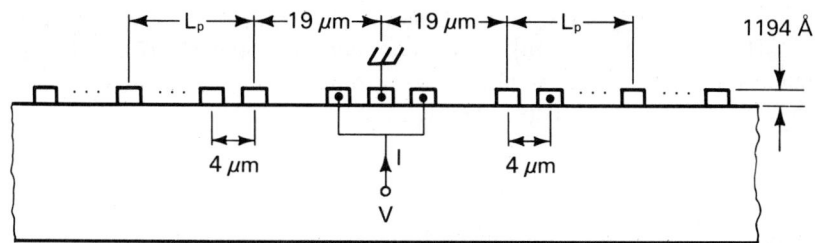

Figure 10.5 Resonator in Example 10.1.

$$f_o = \frac{v_o}{\lambda} = 394.8 \text{ MHz}$$

Solution
From Example 6.3 we have

$$r = -j\, 0.67 \frac{h}{\lambda} = -j\, (0.01)$$

where r is the reflection coefficient of a single strip referenced to its center. Hence we choose the positive signs in Eqs. 10.4a and 10.4b. Equations 10.9a and 10.9b then read

$$L_1 = \frac{m_1 \lambda}{2} + \frac{3\lambda}{8}$$

$$L_2 = \frac{m_2 \lambda}{2} + \frac{3\lambda}{8}$$

m_1 and m_2 being integers. Also,

$$L_p = 1/4|r| = 25\,\lambda \qquad (6.7)$$

$$L_2 = L_1 = L_1' + L_p = 27\frac{3}{8}\lambda$$

Thus Eqs. 10.9a and 10.9b are satisfied. Equation 10.8 is also satisfied since $2(L_1 + L_2) = 109.5\lambda$. This gives us the basic rule for positioning the IDT: The center-to-center distance from the IDT to the strip is $\frac{m\lambda}{2} + \frac{\lambda}{8}$ if the phase is $+\frac{\pi}{2}$ and $\frac{m\lambda}{2} + \frac{3\lambda}{8}$ if it is $-\frac{\pi}{2}$. This is also evident from Fig. 6.7 (Example 6.2), noting that the IDT should have its electrodes positioned at the potential maxima for maximum coupling.

The center frequency of the resonator is approximately 394.75 MHz (v_o = 3158 m/s, λ = 8 μm); however, there is some shift in the center frequency because the presence of the grooved array modifies the velocity of the SAW underneath it (see Section 6.2).

$$\Gamma = \tanh N|r| \tag{6.4c}$$

so that

$$\frac{1-\Gamma}{1+\Gamma} = e^{-2N|r|} = 0.135$$

For the IDT,

$$\mu(f_c) = 0.8jK^2 \cdot 4 \qquad \text{(Example 4.5)}$$

$$G = 2|\mu|^2 y_o \frac{W}{\lambda} \tag{4.21}$$

$$= 2.5 \times 10^{-4} \text{ mmho}$$

$$= \frac{1}{4} \text{ k}\Omega$$

Hence

$$R = 0.135 \times 4 \text{ k}\Omega \tag{10.13a}$$

$$= 541.1 \text{ }\Omega$$

$$Q = 722$$

Note that if N were four times larger (= 400) we would get $R = 4 \times 10^{-4} R_o$ and $Q = 245725$. In practice, of course, we never get Q's this high, because of the propagation losses and any losses due to bulk mode conversion. As a result, $1 - \Gamma$ is always higher than the theoretical value of $2e^{-2N|r|}$ used in this example. If there is a propagation loss of α nepers/wavelength, Γ is reduced by exp $(-\alpha L/\lambda)$, L being the round trip distance from the reflector. As a result,

$$1 - \Gamma \simeq e^{-2N|r|} + \alpha \frac{L}{\lambda}$$

10 RESONATORS

This increase in $1 - \Gamma$ causes an increase in the series resistance R and a lowering of the Q.

10.3. Two-Port Resonator

To make a one-port resonator into a two-port resonator we merely add another IDT to the cavity (Fig. 10.2a). Our objective now is to eliminate all SAW variables so as to get a two-port admittance matrix for the device.

$$\begin{Bmatrix} I_1 \\ I_2 \end{Bmatrix} = \begin{bmatrix} Y_{11} & Y_{12} \\ Y_{21} & Y_{22} \end{bmatrix} \begin{Bmatrix} V_1 \\ V_2 \end{Bmatrix} \tag{10.15}$$

We can get Y_{11} and Y_{21} rather easily using the results of the preceding section. Note that

$$Y_{11} = \frac{I_1}{V_1}\bigg|_{V_2 = 0} \qquad Y_{21} = \frac{I_2}{V_1}\bigg|_{V_2 = 0} \tag{10.16}$$

If $V_2 = 0$, then IDT2 does not generate any waves. So it does not disturb the standing waves set up in the cavity by IDT1, which we analyzed in the preceding section. Hence

$$I_2 = -g_{m2}\left[\phi_2^+ P + \frac{\phi_2^-}{P}\right] \tag{10.17}$$

where g_{m2} is the receiver response function of IDT2 and P is a phase factor that shifts the reference plane for $\phi_2^\pm$ from the center of IDT1 to that of IDT2:

$$P = e^{-j2\pi L_o/\lambda} \tag{10.18}$$

where L_o is the center-to-center distance from IDT1 to IDT2.

Using Eqs. 10.2b and 10.5b in Eq. 10.17, we get

$$Y_{21} = G_{21}\frac{(P + \Gamma_2/P)(1 + \Gamma_1)}{1 - \Gamma_1\Gamma_2} \tag{10.19}$$

where $G_{21} = -g_{m2}\mu_1 = G$ if IDT1 and IDT2 are identical, so that $g_{m2} = g_{m1}$. Y_{11} is simply the one-port admittance we calculated in the last preceding section (Eq. 10.7a).

$$Y_{11} = G \frac{(1 + \Gamma_1)(1 + \Gamma_2)}{1 - \Gamma_1 \Gamma_2} \tag{10.20a}$$

$$Y_{21} = G \frac{P(1 + \Gamma_1)(1 + \Gamma_2/P^2)}{1 - \Gamma_1 \Gamma_2} \tag{10.20b}$$

where we have assumed that $\mu_1 = \mu_2 = \mu$, $g_{m1} = g_{m2} = g_m$, and $G = -\mu g_m$. Similarly, if we short IDT1 and analyze the waves set up by IDT2, we get

$$Y_{12} = Y_{21} \tag{10.20c}$$

$$Y_{22} = Y_{11} \tag{10.20d}$$

It is evident from Eq. 10.20 that $\dfrac{\Gamma_2}{P_2} \simeq 1$ results in a maximum of transconductance Y_{21}. Since $\Gamma_2 \simeq 1$ for Y_{11} to be maximum (see Section 10.2), this requires that $P^2 = 1$; physically, it means that IDT1 and IDT2 should be separated by an integer number of half-wavelengths so that $P = \pm 1$. Hence around the center frequency

$$Y_{21} = Y_{12} = \pm Y_{11}$$

and Y_{11} and Y_{22} are the same as the one-port admittance Y discussed in the preceding section. This justifies the equivalent circuit of Fig. 10.2b, the series resonant arm being exactly the same as that in the one-port equivalent circuit of Fig. 10.1b. As usual, the transducer capacitances were neglected in our discussion and have been inserted in the equivalent circuit. The $\pm$ sign for Y_{12} and Y_{21} is taken care of by reversing the polarity of the output.

Example 10.2
Suppose that the one-port resonator of Example 10.1 were used as a two-port resonator using another identical IDT. What is the loaded Q and the insertion loss if the resonator is driven with equal source and load resistances R_o for (a) $R_o = 10\ \Omega$; (b) $R_o = 200\ \Omega$?

Solution
The loaded resonator has the equivalent circuit shown in Fig. 10.6, with $C_T = 0.035$ pF (see Example 4.17). The loaded Q_L is related to the unloaded Q_u by

10 RESONATORS

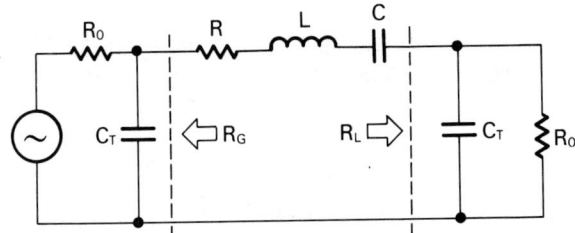

Figure 10.6 Loaded resonator equivalent circuit for Example 10.2.

$$Q_L = Q_u \frac{R}{R + R_L + R_G}$$

where R_L and R_G are the equivalent resistances looking into the load and generator, respectively. Also,

$$\text{IL} = -20 \log \frac{R_G + R_L}{R_G + R_L + R}$$

In this example

$$R_L = R_G = \frac{R_o}{1 + 4\pi^2 f_o^2 C_T^2 R_o^2}$$

$$= R_0 \text{ (approx.)}$$

Hence

$$Q_L = 696 \qquad [\text{part (a)}]$$
$$= 415 \qquad [\text{part (b)}]$$
$$\text{IL} = 29 \text{ dB} \qquad [\text{part (a)}]$$
$$= 7.4 \text{ dB} \qquad [\text{part (b)}]$$

This problem illustrates the trade-off between loaded Q_L and the insertion loss.

10.4 Illustrative Example of Resonator Design[*]

Specifications. To be used in baseband CATV converters to stabilize the second local oscillator.

---

[*]Courtesy of RF Monolithics, Inc.

    Center frequency   567.0 MHz
    Low insertion loss   7 dB typical
    High unloaded Q 9000 typical

Configuration. The substrate used is 34°-rotated y-cut quartz (x propagating direction) with the following design parameters:

1. Reflectors consist of 260 shorted quarter-wavelength stripes.
2. Transducers consist of 110 quarter-wavelength stripes and are apodized to match the lowest propagating-mode energy distribution.
3. Beamwidth is 140 wavelengths.
4. Spacing between transducer/reflector arrays is 11.75 wavelengths.

Measured Performance. Figure 10.7a show's the two-port transfer characteristic. Table 10.1 gives the electrical characteristics measured at 25°C (unless otherwise noted). Figure 10.8 shown the frequency variation with temperature.

TABLE 10.1. Electrical Characteristics of SAW Resonator

Characteristic	Symbol	Minimum	Typical	Maximum	Unit
Center frequency[a]	f_c	566.93	567.0	567.13	MHz
Unloaded Q	Q_u	4000	9000	—	—
Insertion loss[a]	IL	—	7	12.5	dB
Spurious responses	—	-8	—	—	dB
Phase slope at f_c[a]	—	-2°/100 kHz	—	—	—
Power dissipation[b]	P_0	—	—	+10	dBm
Turnover temperature (see temperature curve)	T_o	47.0	62.0	77.0	°C
Aging first year[c]	—	—	—	10	ppm
DC breakdown voltage	V_{Br}	30	—	—	VDC
Input/output capacitance (Fig. 10.6)	C_T	1.6	1.9	2.2	pF
Series resistance (Fig. 10.6)	R	—	125	—	Ω
Series capacitance (Fig. 10.6)	C	—	2.94×10^{-3}	—	pF
Series inductance (Fig. 10.6)	L	—	2.67	—	mH

[a] Measured unmatched in a 50-Ω system.

[b] Matched power to one port, other port shorted to ground with $T_c = 67°C$.
Power degradation threshold — all specifications met after application of +15 dBm under matched conditions above for 5 min at 25°C.

[c] Change in device f_c with time; f_c increases as a logarithmic function of time when operated at a fixed ambient temperature.

10 RESONATORS

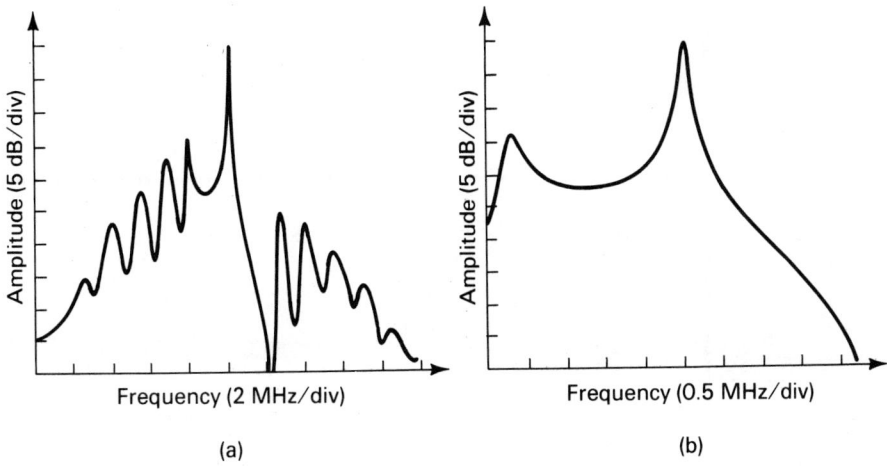

Figure 10.7 Typical resonator transfer characteristic: (a) 2 MHz/div frequency scale; (b) 0.5 MHz/div frequency scale.

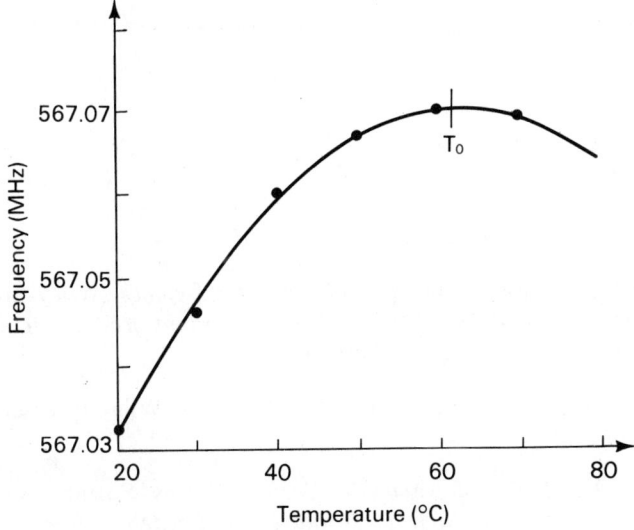

Figure 10.8 Typical frequency versus temperature.

REFERENCES

General

1. B. A. Auld, *Acoustic Fields and Waves in Solids,* Vols. I and II, Wiley-Interscience, New York, 1973. *This is an excellent textbook on acoustic fields that is suitable at an advanced graduate level. The author has learned a lot from this book.*

2. V. M. Ristic, *Principles of Acoustic Devices,* Wiley-Interscience, New York, 1983. *This is also an excellent graduate-level text with more device emphasis than Ref. 1.*

3. H. Matthews, ed., *Surface Wave Filters,* Wiley-Interscience, New York, 1977.

4. A. A. Oliner, ed., *Acoustic Surface Waves,* Springer-Verlag, New York, 1978. *Both Refs. 3 and 4 are excellent references on SAW device design; many different aspects have been covered in individual chapters, each written by an expert in the fields.*

Another general reference on acoustic waves is a series of volumes:

5. W. P. Mason and R. N. Thurston, eds., *Physical Acoustics,* Academic Press, New York, 1964 – (many volumes).

REFERENCES

6. M. Lewis et al., "Recent Developments in SAW Devices," *Proc. IEEE, 131,* Part A, No. 4 (June 1984). *This is a recent review article on SAW devices.*

   Some of the other review articles that have been published in the past are:

7. R. M. White, *Proc. IEEE, 58,* 1238 − 1276 (1970).

8. J. D. Maines and E. G. S. Paige, *Proc. IEEE Rev., 120,* 1078 − 1110 (1973).

9. M. G. Holland and L.T. Claiborne, *Proc. IEEE, 62,* No. 5, 582 − 611 (1974).

10. Special Issue on Microwave Acoustics, issued jointly by *IEEE Trans. Microwave Theory Tech., MTT-27,* No. 11 (1969); *IEEE Trans. Son. Ultrason., SU-16* (1969).

11. Special Issue on Microwave Acoustic Signal Processing, issued jointly by *IEEE Trans. Microwave Theory Tech., MTT-21,* No. 4 (1973), *IEEE Trans. Son. Ultrason., SU-20* (1973).

12. Special Issue on Surface Acoustic Wave Devices and Applications, *Proc. IEEE, 64,* No. 5 (1976).

Chapter 1

1.1. H. E. Kallman, *Proc. IRE, 28,* 302 (1940).

1.2. J. H. McClellan, T. W. Parks, and L. R. Rabiner, *IEEE Trans. Audio Electroacoust., AU-21,* 506 (1973).

   The design of digital filters (both recursive and non-recursive) is discussed in:

1.3. A. Oppenheim and R. Schafer, "Digital Signal Processing," Prentice-Hall, New Jersey, 1975.

   Transversal filters in different frequency ranges are also implemented using charge coupled devices (CCD) and magnetostatic waves. These are discussed in:

1.4. C. H. Sequin and M. F. Tompsett, "Charge Transfer Devices," Suppl. 8 in *Advances in Electronics and Electron Physics,* L. Marton, ed., Academic Press, New York, 1975

1.5. B. A. Auld, *Applied Solid State Science,* Vol. 2, Academic Press, New York, 1981.

REFERENCES

An interesting concept that may be used to implement transversal filters is that of using the electrical potential associated with a SAW to transport charge in a CCD-like configuration.

1.6. S. D. Gaalema, R. J. Schwartz, and R. L. Gunshor, *Appl. Phys. Lett.*, *29*, 82 (1976).

1.7. M. J. Hoskins, H. Morkoc, and B. J. Hunsinger, *Appl. Phys. Lett.*, *41* (1982).

Chapter 2

Bulk Acoustic Waves and their similarities with transmission lines are discussed in Ref. 1 (Chap. 6, Vol. I). Piezoelectric bulk wave transducers and the Mason model are discussed in Ref. 2 (Chap. 5). The normal mode theory is discussed extensively in Ref. 1 (Chap. 6 of Vol. I and Chaps. 10 and 12 of Vol. II). Bulk wave velocities for various cuts and orientation of different substrates are given in:

2.1. A. J. Slobodnik, Jr., R. T. Delmonico, and E. D. Conway, *Microwave Acoustics Handbook,* Vol. 3, Rome Air Development Center, RADC-TR-80-188, 1980.

Chapter 3

Details on the nature and properties of SAW are discussed in several articles by G. W. Farnell such as:

3.1. Ref. 3, Chap. 1.

3.2. Ref. 4, Chap. 2.

3.3. Ref. 5, Vol. 6, p. 109, and Vol. 9, p. 35.

An invaluable reference for the calculated properties of surface acoustic waves on numerous substrates, cuts and orientation:

3.4. A. J. Slobodnik, Jr., R. T. Delmonico, and E. D. Conway, *Microwave Acoustics Handbook,* Air Force Cambridge Research Laboratories, Bedford, Mass., AFCRL-TR-73-0597, Vol. 1, 1973, and AFCRL-TR-74-0536, Vol. 2, 1974.

Another interesting reference on materials for SAW devices is:

3.5. A. J. Slobodnik, Jr., *IEEE Trans. Son. Ultrason., SU-20,* 315 (1973).

REFERENCES

Chapter 4

The number of references on interdigital transducers is truly voluminous. Each of Refs. 1 to 4 has a chapter devoted to the subject with many references. The basic approach we have adopted is a simplified version of the normal mode theory discussed in Ref. 1 (Chap. 10, Vol. II) and also in:

4.1. B. A. Auld and G. S. Kino, *IEEE Trans. Electron. Devices, ED-18,* 898 (1971).

The splitting of the response into element factor and array factor is described in:

4.2. B. Lewis, P. M. Jordan, R. F. Milsom, and D. P. Morgan, Proc. 1978 IEEE Ultrasonics Symposium, IEEE Cat. No. 78CH1344-1SU, 709 (1978); also B. J. Hunsinger and S. Datta, same volume, p. 705.

4.3. S. Datta, B. J. Hunsinger, and D. C. Malocha, *IEEE Trans. Son. Ultrason., SU-26,* 235 (1979); also *SU-27,* 42 (1980).

A reference widely used by SAW filter designers is:

4.4. C. S. Hartmann, D. T. Bell, and R. C. Rosenfeld, *IEEE Trans. Microwave Theory Tech., MTT-21,* 162 (1973).

The design principles involved in withdrawal weighting are discussed in:

4.5. C. S. Hartman, *Proc. IEEE Ultrason. Symp.,* 423 (1973).

4.6. K. R. Laker, E. Cohen, and A. J. Slobodnik, *Proc. IEEE Ultrason. Symp.,* 317 (1976).

4.7. K. R. Laker, E. Cohen, T. L. Szabo, and J. A. Pustaver, *IEEE Trans. Circuits Syst.,* CAS-25, 241 (1978).

4.8. M. Yamaguchi, K. Y. Hashimoto, and H. Kogo, *IEEE Trans. Son. Ultrason., SU-26,* 53 (1979).

Chapter 5

Multistrip couplers were first reported in:

5.1. F. G. Marshall and E. G. S. Paige, *Electron. Lett.,* 7, 460 (1971).

5.2. F. G. Marshall, C. O. Newton, and E. G.S. Paige, *IEEE Trans. Son. Ultrason., SU-20,* 124 (1973).

The coupled mode approach adopted in this book is described in:

5.3. S. Datta and B.J. Hunsinger *J. Appl. Phys.*, *49*, 3769 (1978).

Some of the other interesting references are:

5.4. K. Blotekjaer, K. A. Ingebrigtsen, and H. Skeie, *IEEE Trans. Electron. Devices, ED-20*, 1139 (1973).

5.5. M. Feldmann and J. Henaff, *Proc. IEEE Ultrason. Symp.*, 686 (1977).

5.6. G. S. Kino and W. R. Shreve, *J. Appl. Phys.*, *44*, 3960 (1973).

Chapter 6

The analysis of reflector gratings using repetitively mismatched transmission is described in many references such as:

6.1. E. K. Sittig and G. A. Coquin, *IEEE Trans. Son. Ultrason., SU-15*, 111 (1968).

Experimental results for reflection per strip for various strip materials are described in:

6.2. C. Dunnrowicz, F. Sandy, and T. Parker, *Proc. IEEE Ultrason. Symp.*, 386 (1976).

Reflection due to piezoelectric effects is discussed in:

6.3. T. Aoki and K. A. Ingebrigtsen, *IEEE Trans. Son. Ultrason., SU-24*, 167 (1977).

6.4. S. Datta and B. J. Hunsinger, *J. Appl. Phys.*, *51*, 4817 (1980).

Reflection due to the first-order mechanical effect is described in

6.5. B. K. Sinha and H. F. Tiersten, *J. Appl. Phys.*, *47*, 2824 (1976).

6.6. D. A. Simmons, *J. Acoust. Soc. Am.*, *63*, 1292 (1978).

6.7. S. Datta and B. J. Hunsinger, *J. Appl. Phys.*, *50*, 5661 (1979).

Second-order mechanical effects are discussed in:

6.8. H. Shimizu and M. Takeuchi, *Proc. IEEE Ultrason. Symp.*, 667 (1979).

6.9. S. Datta and B. J. Hunsinger, *IEEE Trans. Son. Ultrason., SU-27*, 333 (1980); also *Proc. IEEE Ultrason. Symp.*, 673 (1979).

REFERENCES

The results for the various types of reflection are summarized for easy reference in:

6.10. B. J. Hunsinger, *Final Technical Report to Rome Air Development Center*, RADC-TR-81-173, 1981.

Chapter 7

7.1. G. S. Kino and T. M. Reeder, *IEEE Trans. Electron. Devices*, ED-18, 909 (1971).

7.2. See Chap. 12, p. 293, in Ref. 1.

An excellent nonmathematical explanation of SAW amplification by drifting carriers is given in:

7.3. R. Adler, *IEEE Trans. Son. Ultrason.*, SU-18, 115 (1971).

7.4. Ref. 1, p. 286, Vol. II.

Chapter 8

8.1. R. V. Schmidt and L. A. Coldren, *IEEE Trans. Son. Ultrason.*, SU-22, 115 (1975).

8.2. L. A. Coldren and D. H. Smithgall, *IEEE Trans. Son. Ultrason.*, SU-22, 123 (1975).

Chapter 9

Diffraction effects in SAW devices are discussed in:

9.1. T. L. Szabo and A. J. Slobodnik, *IEEE Trans. Son. Ultrason.*, SU-21, 114 (1974); SU-20, 240 (1973).

Methods for calculating end effects in interdigital transducers are discussed in:

9.2. C. S. Hartmann, *Proc. IEEE Ultrason. Symp.*, 317 (1976).

9.3. A. L. Lentine, S. Datta, and B. J. Hunsinger, *IEEE Trans. Son. Ultrason.*, SU-26, 53 (1979).

9.4. R. S. Wagers, *Proc. IEEE Ultrason. Symp.*, 536 (1976).

9.5. R. S. Wagers, *IEEE Trans. Son. Ultrason.*, SU-23, 113 (1976).

3 dB multistrip couplers are described in:

9.6. T. I. Browning and F. G. Marshall, *Proc. IEEE Ultrason. Symp.*, 189 (1974).

An interesting case study of a TV IF filter is described in:

9.7. A. J. Devreese and R. Adler, *Proc. IEEE*, 64 (May 1976).

A good reference for the design and analysis of SAW filters is:

9.8. A. J. Slobodnik, Jr., *Surface Acoustic Wave Filters at UHF: Design and Analysis,* Air Force Cambridge Research Laboratories, AFCRL-TR-75-0311, 1975.

Chapter 10

10.1. L. A. Coldren and R. L. Rosenberg, *Proc. IEEE*, 67, 147 (1979).

10.2. D. T. Bell, Jr., and R. C. M. Li, *Proc. IEEE*, 64, 711 (1976).

10.3. E. J. Staples, J. S. Schoenwald, R. C. Rosenfeld, and C. S. Hartmann, *Proc. IEEE Ultrason. Symp.*, 245 (1974).

Recently there has been some interesting work on resonators fabricated on ZnO-on-Silicon. These devices have the potential for integration onto Silicon IC chips.

10.4. S. J. Martin, S. S. Schwartz, R. L. Gunshor, and R. F. Pierret, *J. Appl. Phys.*, 54, 561 (1983).

INDEX

A

Abbreviated subscript notation, 47
Acoustic admittance, 120
　apodized IDTs, 120–22, 130
　unapodized IDTs, 120
Acoustic waves
　orders of magnitude, 15
　plane, 29–67
　surface, 84–138
　uniform plane, 40–46
　velocity, 1
Admittance
　acoustic, 120
　transducer, 120–29
Aliasing error, 13
Alternating-polarity IDT
　example of, 105
Amplifier, surface-wave
　illustration of, 183
Amplitude ripples
　in a bandpass filter, 197
Anisotropic substrate, 192–93
Apodization, 115

Apodized IDT, 120–22, 130
　capacitance of, 126–29
　divided into channels, 133
Apodized transducer, 115
Attenuators, 179–86
　electrical loading, 179–84
　mechanical loading, 184–86

B

Bandpass filters, 124–94
　analysis, 198–214
　basic design procedure, 197–98
　capabilities, 195–97
　center frequency, 196
　examples of filter design, 218–19
　out-of-band rejection in, 197
　power-handling capability, 194–224
　second-order effects, 215–18
Bandwidths, fractional, 197
Beam steering, 82
Bidirectional SAW generation, 106
Brickwall filter, 20

Brickwall response, 21
Bulk-wave generation, 137

C

Capacitance, 126–29
  apodized IDT, 129
  solid-electrode IDT, 128
  split-electrode IDT, 128–29
Capacitor, parallel plate, 90
Center frequency
  of bandpass filter, 196
Charge distribution, 98
Components, SAW device, 84–138
Constitutive relations, 53
Convolution, graphical, 10–12
Convolution theorem of Fourier
  transforms, 9–12
Coupler operation, 143–50
  by a single electrode, 141
Couplers
  multistrip, 139–50
  multitrap, 149–50

D

Debye length, 184
Delay line, 1
  used to provide memory, 14
Delta function stresses, 176
Design
  of narrowband filter, 222–24
  of nonrecursive filters, 20
  procedure, 187–97
  of wideband filter, 218–21
Diagonalizing, 156
Diffraction and propagation losses,
  215
Digital filter, 14–15
  recursive and nonrecursive, 15
Distributed LC-circuit representation,
  30–33
Dolph-Chebyshev window, 23
Double-electrode IDT
  illustration of, 109

E

Electrical loading
  of attenuators and amplifiers,
    179–84
Electric fields
  in the absence of sources, 60
  in the presence of sources, 60
End effect, 92, 217
Equivalent-circuit model, 86
  Mason model, 29, 54

F

Fast Fourier Transform (FFT), 26–27
FFT (Fast Fourier transform), 26–27
Filter design, examples of, 218–24
  narrowband filter of ST-X quartz,
    222–24
    analysis, 223
    configuration, 222
    design, 223
    measured performance, 224
    specifications, 222
  wideband filter on lithium niobate,
    218–21
    configuration, 218
    design, 219–20
    measured performance, 221
    specifications, 218
Filters
  advantage of transveral filter, 17
  brickwall, 20
  definition, 1
  finite impulse response (FIR), 20
  nonrecursive, 14–20
  nonrecursive digital, 15
  N-tap transversal, equally weighted,
    18
  recursive, 14–20
  recursive digital, 15
  RLC, 1
  SAW, 1
  transversal, 1–28
Finite impulse response (FIR) filters,
  20

Finite time duration, 198
Fourier transforms, 2, 125
   properties, 7
      convolution theorem, 9–12
      shifting theorem, 7–9
   transversal
      analysis and design, 1–28
Fractional bandwidths
   in a bandpass filter, 197
Frequency response
   definition, 2
   relationship between impulse
      response and, 2–7
Frequency response error, 25

G

Graphical convolution, 10–12

H

Hamming window, 22–24
Hilbert transform, 86, 124, 125

I

IDTs (Interdigital Transducers), 15
   alternating-polarity, 105
   apodized, 120–22, 130
   as a distributed source, 91–98
   double-electrode, 109
   solid-electrode, 105
      example of, 127–28
   split-electrode, example of, 128–29
   unapodized, 120
Impedance discontinuity, 174–76
   total, 177–78
Impulse response
   of a bandpass filter, 8–9
   definition, 2
   relationship between frequency
      response and, 2–7
   truncation of, 20–22
In-line Mason model, 54, 137

Interdigital transducer, 15, 84–138
Insertion loss
   in a bandpass filter, 197
Isotropic solids, 52
Isotropic substrate
   and waveguides, 187–92

K

Kallman, H. E., 1

L

Lamé constants, 52
Legendre polynomial, 102
Linear two-port network, 34
Lumped circuit, three-port, 37
Lumped circuit equivalent, 33–36

M

MDC, 82
Mason model
   definition, 29
   in-line, 54
Matrix, scatter (*see* Scatter matrix)
Maxwell's first law, 78
Mechanical electrical loading (MEL)
      reflections, 215
Mechanical loading
   of attenuators and amplifiers,
      184–86
Metallic waveguide, 188
Microwave waveguide, 188
Minimal diffraction cut (MDC), 82
Multistrip beam compressor
   illustration of, 146
Multistrip coupler, 139–50
   coupling of tracks by a single
      electrode, 141
   definition, 139
   design, 140
   overall coupler operation, 143–50
Multitrack nonperiodic couplers,
   149–50

N

Newton's law, 54, 174, 175
Nonrecursive digital filter, 15
Nonrecursive (transversal) filter
  design of, 20
Nonrecursive filters, 14–20
Nonsinusoidal time signals, 5
Normal-mode theory, 59–67
N-tap transversal filter, equally
  weighted, 18
Numerical analysis model, 129–38

O

One-port resonators, 227–35
Out-of-band rejection
  in a bandpass filter, 197

P

Parallel plate capacitor
  example of, 90
Perturbation theory, 171
Piezoelectricity, 52–53
Piezoelectric crystals, 52
Piezoelectric scattering, 173
Piezoelectric solids, 15
  generation of acoustic waves in, 53–59
  surface acoustic waves in, 68–83
Plane acoustic waves, 29–67
  compressional waves, 40–46
  piezoelectricity, 52–54
  shear waves, 46–52
  wave generation in piezoelectric solids, 54–67
    Mason model, 53–59
    normal-mode theory, 59–67
Poisson's equation, 56, 92
Power-handling capability
  of a bandpass filter, 194–224
Principle of reciprocity, 66, 86–91

R

Radiation conductance, 85, 130
Radiation susceptance, 84, 130
Rayleigh wave, 68
Receiver response functions, 86–91
Receiving IDT, 209–14
Reciprocity principle of, 66
  example of, 66–67
Recursive digital filter, 15
Recursive filters, 14–20
Reflection of SAW, 169–78
  transmission lines, 171–73
Reflection regeneration, 215
Reflections, internal, 215
Reflector array, 153–61
  operation, 153–61
  reflection, effective center of, 158
  reflector bandwidth, 157–58
  transmision matrix formulation, 153
Reflectors, 151–78
  definition, 151
  surface wave, 154
Regenerated wave, 215
Resonators, 225–39
  example of resonator design, 237–39
    configuration, 238
    measured performance, 238
    specifications, 237
  one-port, 227–35
  SAW resonator capabilities, 226–27
  two-port, 225, 235–37
Ripple levels
  reduction of, 22
Ripples, amplitude, 197
RLC filter
  circuit diagram, 3
  input signal, 2
  memory, 2
  output signal, 2

S

Sampling theorem, 12–14

Index

SAW (surface acoustic wave), 115
  introductory description, 69
  transmission-line model, 72
SAW, reflection of, 169–78
SAW bandpass filter
  illustration of, 195
SAW delay line, 87
SAW devices, 15
  bandpass filters, 194–224
  low-frequency limit, 25
SAW resonator (table of electrical characteristics), 238
SAW waveguide
  thin-film (illustration of), 188
Scatter matrix, 34
Second-order effects, 215–18
  diffraction and propagation losses, 215–16
  end effects, 217
  internal reflections, 215
  spurious losses, 217–18
  triple-transit echo, 217
Shearing strains
  definition of, 48
Shear waves, 46–52
Shifting theorem of Fourier transforms, 7–9
Sidelobe levels
  reduction of, 22
Single-tap transducers, 102–5
Solid-electrode IDT, 105
  capacitance of, 127–28
  example of, 127–28
Split-electrode IDT
  capacitance of, 128–29
  example of, 128–29
Spurious losses, 217
Strain constant, 81
Stress field, surface-wave, 77
Subscript notation (*see* Abbreviated subscript notation)
Substrate cut
  minimal diffraction cut, (MDC) and propagation direction, 82
Substrates
  in SAW technology, 53

Superposition principle, 98
  illustration of, 99
Surface acoustic waves (SAW), 15
  excited by interdigital transducers, 84–138
  introductory description, 69–72
  substrate cut and orientation, 82–83
  transmission-line model, 72–76
Surface-wave amplifier
  illustration of, 183
Surface waves
  decay, 69
  propagation, 69
Surface-wave stress fields, 77
Surface-wave waveguide, 188–89

T

Thin-film SAW waveguide
  illustration of, 188
Three-port lumped circuit, 37
Total impedance discontinuity, 177–78
Transducer admittance (*see* Admittance, transducer)
Transducers
  apodized, 115
  response functions, 102–20
  single-tap, 102–5
  unidirectional, 112–13
  unweighted, 105–13
  weighted, 113–20
  withdrawal weighted, 113–15
    design procedure, 115
Transition bandwidth
  of a bandpass filter, 196
Transmission line, 29–67, 171–73
  modeled as a distributed *LC* circuit, 30–33
  similarities and differences, 169–78
  wave generation by a current source, 36–40
Transmission matrix formulation, 153
Transmitter response functions, 86–91
  $\mu(f)$, 91–120

Transmitter response functions (*cont.*)
  general theory, 91–102
Transmitting IDT, 198–209
Transversal filters
  advantage of, 17
  Fourier transform properties, 7–14
    convolution theorem, 9–12
    sampling theorem, 12–14
    shifting theorem, 7–9
  frequency response, 2–7
  impulse response, 2–7
  nonrecursive filters, 14–20
    design, 20–28
  recursive filters, 14–20
Transverse and effect, 217
Triple-transit echo, 214, 217
Truncation
  of impulse response, 20–22
Two-port network
  linear, 34
Two-port resonators, 225, 235–37

U

Unapodized IDTs, 120
Unidirectional transducers, 112–13
Uniform plane acoustic waves, 40–46
Unweighted transducers, 105–13

V

Voltage standing-wave ratio, 207

W

Waves, acoustic
  velocity, 1
Wave generation
  by a current source, 36–40
Waveguide
  anisotropic substrate, 192–93
  isotropic substrate, 187–92
  metallic, 188
  microwave, 188
  zigzag wave, 192
Weighted transducers, 113–20
Windowing functions, 20–24
  Dolph-Chebyshev, 23
  Hamming, 22–24
  rectangular, 21–24
Windowing technique
  causing error in frequency response, 25
Withdrawal weighted transducers, 113–15
  design procedure, 115

Y

Y-cut lithium niobate
  illustration of, 82

Z

Zigzag-wave waveguide, 192